MASONRY CONSTRUCTION: THE TROWEL WORKER'S BIBLE

Build solid, professional-looking masonry structures . . . *the first time, every time!*

BY J. M. NICKEY

 TAB BOOKS Inc.

BLUE RIDGE SUMMIT, PA. 17214

FIRST EDITION

FIRST PRINTING

Copyright © 1982 by TAB BOOKS Inc.

Printed in the United States of America

Reproduction or publication of the content in any manner, without express permission of the publisher, is prohibited. No liability is assumed with respect to the use of the information herein.

Library of Congress Cataloging in Publication Data

Nickey, J.M.
    Masonry construction.

    Includes index.
    1. Masonry—Amateurs' manuals. I. Title.
TH5313.N5   693.'.1   81-18328
ISBN 0-8306-0062-0   AACR2
ISBN 0-8306-1280-7 (pbk.)

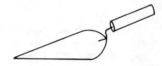

Contents

Introduction

This book will make construction with masonry units more familiar to everyone. The uses of trowels and other necessary tools are fully explained. The various units of masonry are presented as to kind, size, and shape. The practical methods used when setting and laying up masonry structures, although varied, are discussed and illustrated. The numerous kinds of mortar mixes and their uses are covered.

From the time man first used natural units of construction, the art of building with masonry units, natural and fabricated, has advanced to present day methods. A mortar mix is required whenever a trowel is used in today's masonry.

There are many methods, centuries old, that are still used when doing masonwork. Contemporary materials, high quality cements, limes, and power tools will be used for many years to come.

Comprehension and proficiency in the art of masonry construction is your desire and the book's purpose. If you read and practice the methods discussed in this book, you will become proficient in the art of masonwork. I have written this book for the novice, the apprentice, the journeyman, and the contractor.

To all workers who want to pick up trowels and build big and small structures.

Other TAB books by the author:

No. 1226 *The Stoneworker's Bible*

Chapter 1

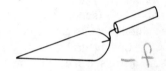

Masonry

The *mason* and the various masonry units such as stone, brick, tile (structural and decorative), and block, including plaster, stucco, mortar, and cement finish are all considered *masonry*. There are stonemasons, brickmasons, tile and block masons, and also plastering, stuccoing, cement finishing, and tucking masons. A mason is a person competent to a certain degree, depending on his knowledge and practical ability, in whichever kind of masonry he works. He may have started as a novice, become an apprentice, and finally worked up to journeyman status. He may be a capable and competent mason in one or more kinds of masonwork and construction.

A stonemason, a brickmason, or any other mason should be and generally is capable of handling a trowel and the various tools in the masonwork he does. He should be capable of shaping, cutting, handling, setting, or laying the units of masonry construction he puts together. He is knowledgeable and competent in mixing and handling mortar. He is capable of cleaning and washing his finished masonwork.

MASONRY UNITS

The first builders used natural masonry units. Masonry units have many sizes and are of various materials. Stone is a natural unit. As construction techniques advanced, other masonry units were introduced. Fabricated masonry units are durable, beautiful, and serve as construction units. In many cases man-made masonry units have outlasted nature's units.

There are two main groups of fabricated masonry units. One group is made from an *aggregate* and a *binder*. The other, made from clays, are coalesced into solid units by heat and fusion. These can be fired to a *vitreous* and semivitreous state. Both groups serve their various purposes well.

STONE

Some *stone* is useless as a masonry unit. Stone which originates from *igneous, metamorphic*, and *sedimentary* stone can be used as masonry units. These stones include *limestone, granite, sandstone* (sedimentary and stratified), *marble*, and *slate*. Other natural stone is also used.

Both natural and artificial stone units and stonework methods are discussed in depth in my book *The Stoneworker's Bible*. Stone is shown in combination with other masonry units.

In the making of artificial masonry stone units, an aggregate and a binder are used. The aggregate can be natural or fabricated. There are various binder mixes. Cements are the most common.

STONEWORK AND BRICKWORK

Neither *Stonework* nor brickwork are discussed in depth here except in combination with other masonwork. These subjects are covered in detail in other books such as my books *The Stoneworker's Bible* and *Nickey's Brickwork Construction* (published by the Dorrance Company).

BLOCKWORK

Blockwork will be disclosed in depth in this book. There are many kinds of structural and decorative blockwork. Blockwork could be discussed as tilework, but it will be blockwork in this book.

GLASS BLOCK

Glass block can be purchased in various sizes. Take care when laying glass blocks. These blocks expand and contract in all temperatures and require expansion room when laid. Some of the blocks have inside and outside faces. One side is opaque and the other can be seen through. Some blocks are made to reflect heat, and some have directional lighting qualities. These blocks come with different shaped faces and sizes.

GYPSUM TILE

Tile made from *gypsum* is used in interior, nonbearing walls as a partition and inside veneer. It is fire-resistant.

Gypsum tile can be cut with a handsaw or power saw. There is little waste. It comes in 2″ solids (2″ × 16″ × 32″) or 3″, 4″, and 6″ thicknesses in hollow tile.

MAGNESIA BLOCK

Magnesia block is a good sound acoustical block. It requires special handling when it is used as a veneer for sound resistance.

STRUCTURAL TILE

There are two kinds of *tile—structural* tile and *decorative or floor and wall* tile. Structural tile withstands diagonal and horizontal stress or strain to a certain extent. Structural tile may hold a live and dead load according to its crushing strength. Floor and wall tile is more a decorative unit coating which withstands the elements and is pleasing to the eye.

Structural tile is made from clay or other materials. Structural tile comes in various shapes and sizes. There is special purpose tile as well as common tile.

FLOOR AND WALL TILES

Floor tile is different from structural tile in that floor tile, or wall tile, is used more as a veneer. Wall tile is not required to hold up a bearing weight. The sizes of floor tile and wall tile can vary in breadth, width, and thickness. Floor tile and wall tile can be purchased in sizes divisible by 2″. There are small tiles less than 2″ in breadth, width, or depth (thickness). There are 4 × 4 tiles, 6 × 6 tiles, 8 × 8 tiles, 10 × 10 tiles, and so on in even larger sizes divisible by 2″. The width may not in some tiles be the same size as the breadth.

Tile may consist of stone, clay, concrete, or other fabricated materials. There are also various textured tiles, some with figures and even pictures on them. Tile generally is glazed with extreme heat, or the biscuit is coated with colorful enamel before heating (this heating process pertains to clay tile).

Tile can be cut in various shapes and sizes. Hand tools and power tools are used.

There are inside and outside floor tile and wall tile. The tile used inside sometimes is subject to moisture. It should not be used outside. The outside tile is generally capable of withstanding the elements. It can preserve and outlast marble. This outside tile is generally glazed.

TUCKING

Tucking is the term applied to filling mortar joints on new and old masonwork. Tucking requires tools comfortable to the work of the tucking trade. These tools are of various sizes and shapes. Some of them are used for cleaning out the old joint mortar and masonwork that needs repairing.

PLASTERING

Plastering is an art. A good plastermason considers himself an artist. Plastering requires appropriate touches with respect to time in order to do the best job of application. There are various kinds of plaster finishes and plasterwork presentations.

STUCCO

Stucco is generally considered as an outside coat applied to a firm stable base. There are many kinds of stucco. Stucco is applied by hand or machine.

CEMENT FINISHING

There are various tools required by the cement mason to perform masonry finishing and concrete repairing. These tools come in many shapes, sizes, and widths, and lengths.

CLEANING MASONWORK

Cleaning masonwork requires good judgment and an understanding of the masonry construction to be cleaned. When cleaning masonwork, care must be taken not to damage the masonry finish joints and the masonry units.

Chapter 2

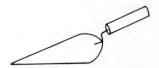

Tools and Equipment

The tools and equipment used when building with the various masonry units are discussed in this chapter and throughout the book. Their many uses are illustrated.

The methods used in handling these tools and laying the units become individually unique with different masons. The masons formulate their own methods. It is the end result that counts.

TROWELS

There are several kinds of *trowels* used when doing masonwork. Some are better adapted and more practical to use than others when building with certain masonry units.

High carbon steel blades are used when making a good trowel. They are ground to shape and checked for flaws. These are short and high posts, depending on the make of trowel, used on which to put either a long or short handle. Handles are either hard wood with a steel ferrule or plastic. The short post trowels have a low hang and

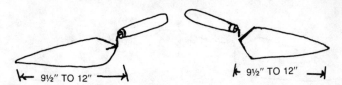

Fig. 2-1. Rounded narrow heel trowels.

generally a stubby handle. The high post trowels have a higher hang and generally longer (6″) handles.

Trowel blades are of different sizes, from 3″ to 16″ in length and from ⅜″ to 6″ in width depending on the use. They come in various shapes.

The rounded narrow heel trowel is preferred by many masons, especially when doing veneer work. The blade is less than 5″ wide at the blade heel and comes in blade lengths from 9½″ to 12″ or more (Fig. 2-1). A semirounded pointed and wide heel trowel blade is used by many masons when laying tile and block (Fig. 2-2).

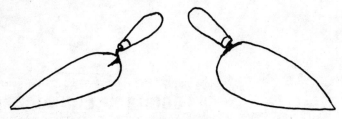

Fig. 2-2. Semirounded pointed and wide heel trowels.

The British trowels are liked by masons. The post is stubby and so is the handle. The trowels are available in many blade widths and lengths (Fig. 2-3).

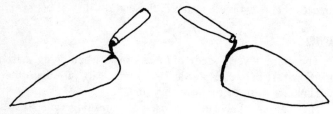

Fig. 2-3. British trowels.

The wide pointed heel trowel is also liked by many masons. Its blade heel generally is 5½″ wide and comes in lengths of 10″ to more than 12″ long (Fig. 2-4).

Fig. 2-4. Wide pointed heel trowels.

There are trowels shaped for pointing, buttering, gauging, and joint and margin work. The *pointing* trowels come in different sizes and are used for pointing joints, principally on old work (Fig. 2-5).

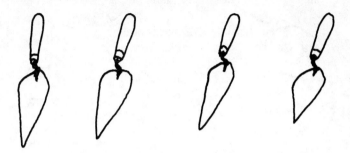

Fig. 2-5. Pointing trowels.

The *margin* trowels are handy ones for a mason to have. They come in widths from 1½″ to 2″ and lengths from 5″ to 8″. They should have flexible blades even though they are taper ground. Generally, the posts are high. A long handled one is easiest to use (Fig. 2-6).

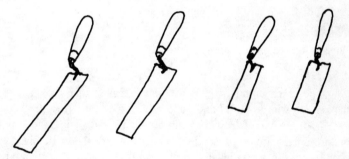

Fig. 2-6. Margin trowels.

The *buttering* trowel is used when laying wall or floor tile (Fig. 2-7). It is a wide, short-bladed trowel with a width of approximately 4¾″ and a length of 6″ or 7″. The tapered gauging trowel comes with a 3″ by 7″ blade (Fig. 2-8).

Caulking and *tucking* trowels can be purchased in widths of ¼″, ⅜″, ½″ ⅝″, ¾″, ⅞″, and 1″. They are generally 7″ long. They come in handy for tucking joints and caulking work. The blades should be flexible, and the post should be high with a long handle. These trowels help reduce knuckle bruises.

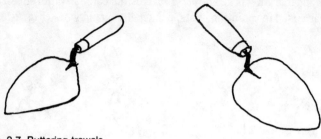

Fig. 2-7. Buttering trowels.

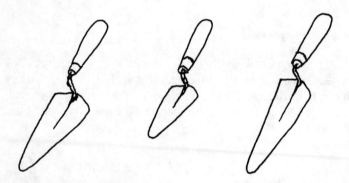

Fig. 2-8. Tapered gauging trowels.

The *cross joint* trowel is a smaller trowel than the gauging trowel. Its width is 1¾″ and generally is 4¾″ long (Fig. 2-9).

Another useful trowel is the *bucket* trowel. It is used when making small batches of mortar in a bucket or small box. Its blade is stiff. Generally, a mason will cut off an old laying trowel and not buy a bucket trowel (Fig. 2-10).

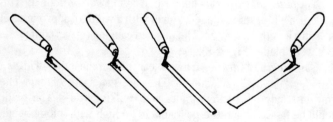

Fig. 2-9. Cross joint trowels.

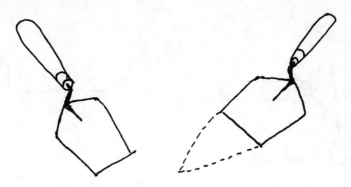

Fig. 2-10. Bucket trowels.

HAMMERS

Hammers are used when building with masonry units. An all-purpose hammer is the *claw* hammer. There are straight claw and curved claw hammers. Their weights can range from 12 to 28 ounces. These claw hammers are used when driving or pulling nails (Fig. 2-11).

Fig. 2-11. Claw hammers.

The mason's *brick* hammer comes in different sizes and weights. The hammers weigh from 12 to 24 ounces. There are different makes. Their shapes are similar, although their heads are of different lengths. The handles are either all metal and covered or wooden (Fig. 2-12).

The *tile layer's* hammer is used when setting wall or floor tile (Fig. 2-13). The *firebrick* hammer is a stubby headed hammer used when scaling and cutting firebricks. Its weight is from 12 to 16 ounces (Fig. 2-14).

The rawhide core hammer has a rawhide filler. This mallet is used when pounding units into place. It will not scar or scratch when used, as metal hammers will. Some mallets have rubber, wood, or lead heads (Fig. 2-15).

Fig. 2-12. Brick hammers.

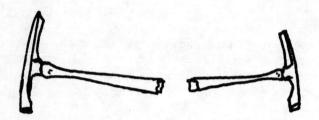

Fig. 2-13. Tile layer's hammers.

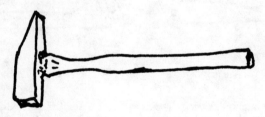

Fig. 2-14. Firebrick hammer.

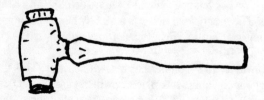

Fig. 2-15. Rawhide core hammer.

There is a hammer called a *scutch*. It has a chisellike blade on both heads that can be replaced when its chisel head wears out (Fig. 2-16).

Fig. 2-16. Scutches.

Another useful hammer is the prospector's *pick* hammer. Its weight is from 16 to 24 ounces. A small-toothed *brush hammer* is useful (Fig. 2-17).

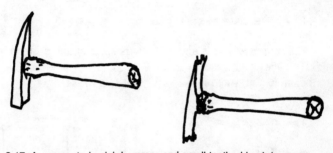

Fig. 2-17. A prospector's pick hammer and small-toothed bush hammer.

JOINTER TOOLS

Joint finishing tools are of various shapes and sizes. The various shaped tools leave different shaped joints when used. A tooled joint not only has a good finished appearance, but withstands the elements of nature much better than an untooled joint. Tooling a joint packs it, makes it less porous and gives it solidity which makes it last longer.

The tools are made of glass, metal, high carbon steel, or stainless steel. The longest jointer tool is called a sledrunner. It is mostly used on blockwork joints (Fig. 2-18).

The *concave* joint is made by a jointer which is available in various sizes to make different size joints. Generally, the size jointer needed is determined by the thickness or breadth of the

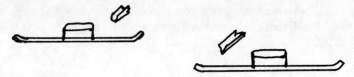

Fig. 2-18. Sledrunners are the longest jointer tools.

mortar joint. The jointer required should be a little wider than the joint to be finished. This helps seal the mortar to both the top and bottom of the mortar joint. Where the joint is recessed, the jointer should be no wider than the mortar joint (Fig. 2-19).

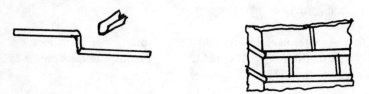

Fig. 2-19. A jointer and a concave mortar joint.

The *convex jointer* is also of various sizes and is used on both ends. It leaves a convex mortar joint (Fig. 2-20). The *V-joint* jointer is a heavier jointer and leaves a V-joint (Fig. 2-21).

Fig. 2-20. A jointer and a convex mortar joint.

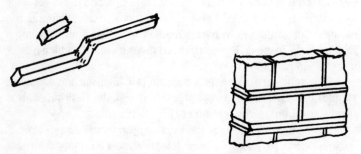

Fig. 2-21. The V-joint jointer and a V-joint.

12

The *rattail* jointer has the rattail end curved and tapered. The other end is straight. It fits recessed concave mortar joints (Fig. 2-22).

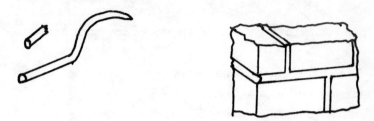

Fig. 2-22. The rattail jointer fits recessed concave mortar joints.

There are two kinds of *raking* jointers used for raking out mortar for a recessed joint (Fig. 2-23). The *projecting square mortar* joint is made with a jointer shaped as in Fig. 2-24.

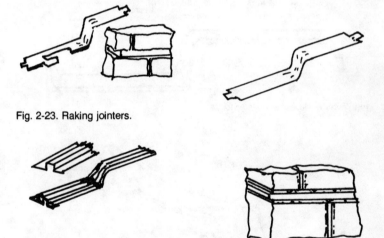

Fig. 2-23. Raking jointers.

Fig. 2-24. A jointer for making a projecting square mortar joint.

The *V-projecting* jointer leaves a V-projecting mortar finish joint (Fig. 2-25). The V-recessed joint is made by the jointer in Fig. 2-26.

The slicker and raker jointers are very handy tools. They have one end wider than the other and can be made in different sizes (Fig. 2-27). The *wheel raking* jointer can be set for depth to rake out mortar joints (Fig. 2-28).

Fig. 2-25. The V-projecting jointer and a V-projecting mortar finish joint.

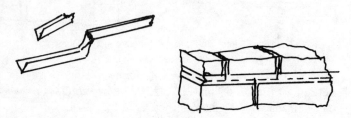

Fig. 2-26. A jointer for making a V-recessed joint.

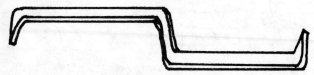

Fig. 2-27. A slicker and raker jointer.

Fig. 2-28. Wheel raking jointer.

Another tool that is used for cleaning off mortar around a joint is the *frenchman* (Fig. 2-29). It is used as a scraping and cleaning tool. The old mortar joint tool is used when raking back or cutting out old mortar to replace it with new mortar (Fig. 2-30).

Fig. 2-29. The frenchman is a cleaning and scraping tool.

Fig. 2-30. Old mortar joint tools are used when raking back or cutting out old mortar.

BRUSHES

There are several kinds of brushes used during masonwork. They have special uses.

The *bricklayer's* brush is used for brushing off spilled and excess mortar after a joint is tooled (Fig. 2-31). The long, thick-

Fig. 2-31. The bricklayer's brush.

haired brush or *splashing* brush is useful for splashing water on the work, especially before tucking a joint (Fig. 2-32).

Fig. 2-32. Splashing brushes.

15

Acid brushes are used for cleaning and washing finished masonwork (Fig. 2-33). Scrub brushes are used for scrubbing finished work. There are several kinds (Fig. 2-34). A wire brush is necessary when washing and scraping down masonwork (Fig. 2-35).

Fig. 2-33. Acid brushes.

Fig. 2-34. Scrub brushes.

Fig. 2-35. Wire brushes.

CHISELS

Chisels of different sizes and shapes are used when building masonwork. The bricklayer's set chisel is used with the hammer (generally the bricklayer's hammer) when cutting block or tile units (Fig. 2-36).

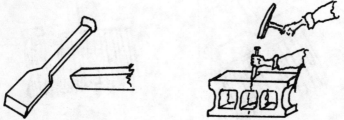

Fig. 2-36. The bricklayer's set chisel is used with a bricklayer's hammer.

16

Another useful chisel is the *wide blade* concrete chisel. This chisel is tempered to withstand hard material. The chisel holds its sharpness during hard usage (Fig. 2-37). There are several points and cold chisels of various sizes and tempers used on tile (Fig. 2-38).

Fig. 2-37. A wide blade concrete chisel.

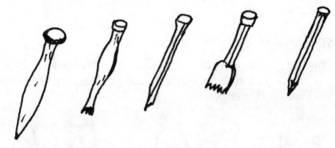

Fig. 2-38. Points and cold chisels are used on tile.

POWER CUTTERS AND DRILLS

There are power tools used for cutting, drilling, chiseling, and scaling construction units. They are powered by either air, electricity, or fuel motors.

The hydraulic-powered cutter is powered by a hydraulic jack. It is pumped with the foot (Fig. 2-39). Electric power drills can handle various bits and sanders (Fig. 2-40).

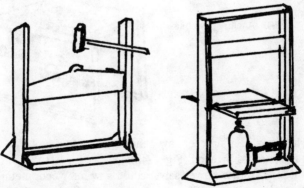

Fig. 2-39. Power cutters.

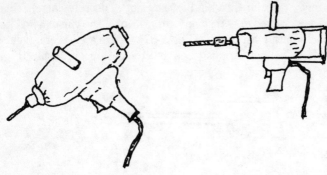

Fig. 2-40. Power drills.

POWER SAWS

Masonry power saws can be obtained to handle blades of many sizes. The diamond blade will outlast many Carborundum abrasive blades (Figs. 2-41 and 2-42).

Fig. 2-41. A power saw and blades.

Fig. 2-42. Another power saw.

The tile setter's power saw generally sits on a tub. The tool cutter is operated by hand (Fig. 2-43).

There are other power saws: cutoff saw for wood, metal, or masonry; close cutting flush to floor saws; Carborundum and

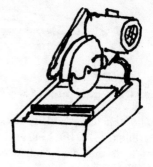

Fig. 2-43. A tile setter's power saw.

diamond blade saws; and handheld, frame-held and movable saws, which can be used for all kinds of cutting, grinding, sanding, and *scarifying*. These saws can be used on all materials, depending on the blades.

OTHER NECESSARY TOOLS

Bolt cutters are used to cut bolts, ⅝″ re-bar, and heavy reinforcing material (Fig. 2-44). Smaller cutters are more handy to use when cutting smaller material (Fig. 2-45).

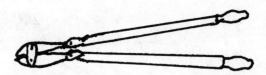

Fig. 2-44. A bolt cutter.

Fig. 2-45. A handy small bolt cutter.

Snips are needed when cutting out flashings (Fig. 2-46). *Tongs* are useful when carrying more than one unit of construction (Fig. 2-47).

Fig. 2-46. Snips are used to cut out flashings.

19

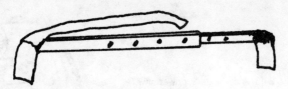

Fig. 2-47. Tongs are used to carry construction units.

Line holdings and *line blocks* are used in place of *line pins* where possible (Fig. 2-48). Line pins, twigs, and masonry lines are necessary tools that have their use when building with masonry units (Fig. 2-49).

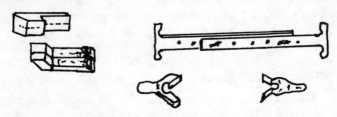

Fig. 2-48. Line holdings and line blocks.

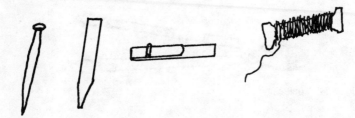

Fig. 2-49. Line pins, twigs, and masonry lines.

Rubber or plastic gloves are generally needed when washing masonwork. Some washing materials burn hands (Fig. 2-50).

Fig. 2-50. Rubber gloves.

20

Fig. 2-51. Dust masks and goggles.

Dust masks and *goggles* are necessities when sawing or working in dusty places (Fig. 2-51).

Caulking guns are used when caulking around openings. Large, powerful ones are used when caulking with grout mortar (Fig. 2-52).

 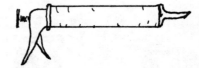

Fig. 2-52. Caulking guns.

LEVELS AND HAWKS

There are various kinds of *levels*. There are horizontal and perpendicular levels of various lengths, *water* levels, and *transit* levels. These levels are used when plumbing and locating grade. They have their own particular purposes (Fig. 2-53). A *hawk* is handy for holding mortar when tucking or applying base material for tile setting (Fig. 2-54).

MEASURING AND MARKING TOOLS

The *folding rule* is marked off in inches and parts of inches. Some rules have elevation markings on them, as well as divisions in feet and inches (Fig. 2-55).

Measuring tapes with various markings on them, even metric markings, can be procured in various lengths up to 100″ long. Some tapes retract themselves, especially the pocket spring retrieving tapes. Tapes are used to measure distances from point to point (Fig. 2-56). *Chalk lines* are used to mark material from point to point (Fig. 2-57). Different kinds of squares are used (Fig. 2-58).

RUBBER BRICKS AND HAULING DEVICES

Rubbing bricks or stones, in various coarsenesses, are used for grinding and smoothing down material (Fig. 2-59). *Wheelbarrows* and *carts* are used for hauling material (Fig. 2-60).

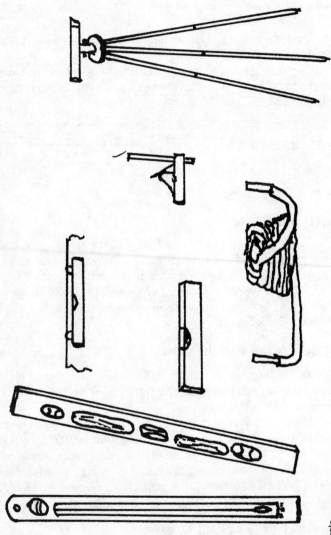

Fig. 2-53. Levels.

22

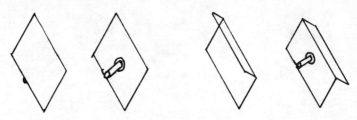

Fig. 2-54. Hawks.

Fig. 2-55. Measuring devices.

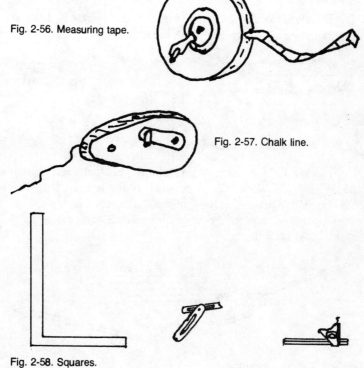

Fig. 2-56. Measuring tape.

Fig. 2-57. Chalk line.

Fig. 2-58. Squares.

Fig. 2-59. Rubbing bricks.

Fig. 2-60. A wheelbarrow and a cart.

MORTAR TOOLS

The *mortar box* is necessary when making mortar by hand. A *water barrrel* and tub are also useful (Fig. 2-61).

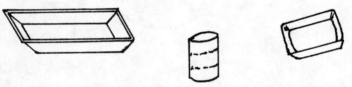

Fig. 2-61. A mortar box, water barrel, and a tub.

The *hoe* and long-handled shovel are tools that every mason should have (Fig. 2-62). *Tubs* or small boxes are needed when making small batches of mortar or grout. The pail is also used for small batches. The *bucket trowel* or *powered drill mixer* can be used for mixing (Fig. 2-63).

Fig. 2-62. A shovel and a hoe.

Fig. 2-63. Tubs and pails are needed when making mortar.

SCREENS AND MIXERS

There are screens of different sizes for screening aggregate from fine to coarse particles (Fig. 2-64). There are several kinds of power mortar mixers. I like the paddle one best (Fig. 2-65).

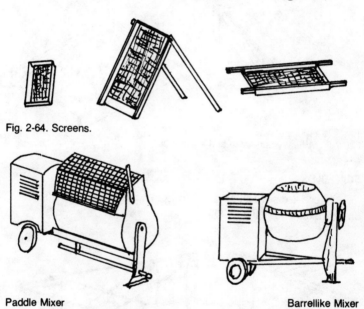

Fig. 2-64. Screens.

Paddle Mixer   Barrellike Mixer

Fig. 2-65. Mortar mixers.

QUANTITY TOOLS

The bottomless cubic foot *measuring box* is used to keep mixes proportionally correct for certain mortars. A pail is also utilized for this purpose. Shovel counts are also used; however, the mix could be out of correct proportions by shovel counting (Fig. 2-66). The heavy-duty *table* or *sandbox* are welcome tools when hand cutting block and structural tile (Fig. 2-67).

Fig. 2-66. A measuring box, shovel, and a pail.

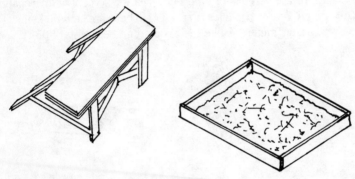

Fig. 2-67. A heavy-duty table and a sandbox.

SCAFFOLDS

There are several kinds of scaffolds. Sawhorses are also used on a masonry job (Fig. 2-68).

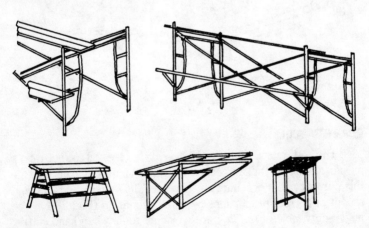

Fig. 2-68. Scaffolds.

HEATERS, WATER TOOLS, AND A PLASTERER'S TROWEL

There are different types of heaters used on masonry jobs to keep the masonry mortar from freezing until set. The fuels used can be oil, gas, electricity, wood, and coal (Fig. 2-69).

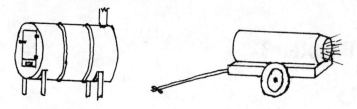

Fig. 2-69. Heaters.

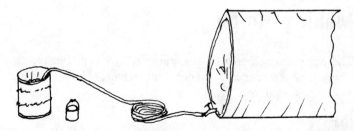

Fig. 2-70. Items used for handling water.

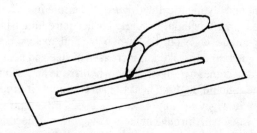

Fig. 2-71. A plasterer's trowel.

A water barrel, hose, and spray nozzles are used for handling water (Fig. 2-70). A plasterer's trowel is used when spreading mortar as a base for setting tile. These trowels vary in size from 3″ × 7½″ to 4½″ × 20″ (Fig. 2-71).

Chapter 3

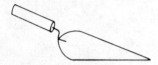

Making Mortar

There are various ingredients incorporated in the different mortars. These ingredients include binders, aggregates, colors, water, and sometimes additives.

LIME

Lime is one ingredient used in some mortar mixes. It is a good binder. In its natural state lime is found in limestone (an *argonite* which is chemically identical with *calcite*). Marble in its best white state makes top grade lime. Properly burned pure lime which has a small percentage of impurities when burned makes quicklime. When this quicklime is slaked, it increases in size and heat. During its slaking it becomes a paste of calcium hydroxide and is known as a *hydrate* lime. Should the lime heat to great activity, it produces fat lime. When the right amount of water is used in the lime's slaking, the putty formed will, through moisture evaporation, become a lime powder known as hydrated lime.

Mortar made from hydrated lime is a binder. It sets under water.

Limestone containing 15 to 25 percent clay is the best stone to use when making lime. Stone such as limestone containing carbonate of lime, with carbonate of magnesia and silica, makes good lime when processed. Calcining or burning such stone at a low temperature forms *clinkers*. These can be pulverized. When slaked with water and used in a mix for mortar, they have the ability to cause the mortar to harden under water.

There are three methods generally used when slaking calcined lime. Lime takes water right at two-thirds of the weight of calcined lime when slaking.

Drowning Method

The *drowning method* of slaking lime to ready it for use is a good one. The calcined lime is put in a flat box to a depth of 6″. Spread it out evenly with a hoe. Add water until the lime is covered well. Put in water at least to two-thirds the weight of the lime. Let the lime be undisturbed until the slaking process is complete. This takes several days. Should the lime be disturbed before its slaking is complete, the result will be lime putty instead of powdered lime. Undisturbed lime in its slaking will after evaporation become powdered lime (Fig. 3-1).

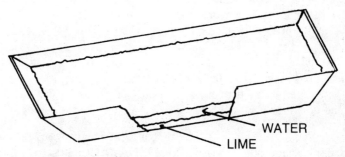

Fig. 3-1. The drowning method of slaking lime.

The Immersion Method

The *immersion method* requires the calcined lime to be put in a porous sack. The sack of lime is then immersed completely in a barrel of water. The lime absorbs enough of the water required for slaking. The sack of lime is then set aside to slake. It will turn to a lime powder after evaporation. This method also takes several days (Fig. 3-2).

Sprinkling Method

Place lime in a box as in the drowning method. Sprinkle with water until all the lime is saturated. After sprinkling, the lime should not be disturbed until the water has slaked the lime and evaporated. No more water should be added, or the lime will turn from powder to lime putty. Lime putty should continuously be kept wet (Fig. 3-3).

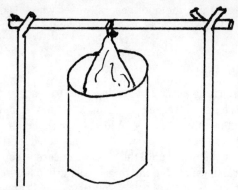

Fig. 3-2. The immersion method of slaking lime.

Slaked lime will keep indefinitely. It should be tightly covered and can be covered with sand to a depth of ½″. Slaked lime can be used after its slaking hydration process is over. Should it be used before the process is complete, the unslaked lime blisters out in the masonry joints to finish its slaking. This ruins the joints. Never add water when lime is in its slaking stage as this stops slaking. Such lime used in mortars later starts to finish its slaking.

Properly slaked calcined lime becomes hydrated lime. This lime in mortar has the property to harden under water.

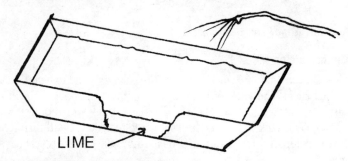

LIME

Fig. 3-3. The sprinkling method of slaking lime.

Lime made from limestone having carbonate of lime with carbonate of magnesia and silica (clay) can make good mortar without an addition of cement. This lime can be made into a lime putty by adding water. The powdered lime or the lime putty can be used for mortar mixes. They can also be used when cement is added as a binder. Cement lime binders make harder mortar. This mortar sustains more weight load than if lime binder is used in the mortar.

CEMENT

There are various cements: *natural* cement, *portland* cement, *Keene's* cement, etc. Cement is one of the best binders used in mortars. It makes the strongest mortar. Some cements have had the clinkers ground to a powder and burned several times to make certain cements. Additives are added to make uniform cement. Testing is carried on continuously when making cements.

Natural Cement

Natural cement is not a strong cement when used in mortars as a binder, unlike portland cement. Natural cements are made from rock having ingredients that vary. Cements made from this rock are low in strength. They should be used only in mortars in which variance in quality does not matter, such as in unbearing partitions and the like.

Portland Cement

Portland cement is made from limestone, shale, and percentages of clay. The quality of this rock is kept uniform by testing, and any variation in the chemical condition is corrected. The materials are calcined (burned) through the kiln, and the clinkers are cooled. The cooled clinkers are pulverized into powdered portland cement. Portland cement as a binder makes a strong mortar which withstands a load weight and pressure.

All cements like limes should be covered and kept in a dry place. Water causes them to set up. If they set up, they are useless as a binder. Dry cement can be shook loose, screened, and used in mortars.

Additives and Binders

Mortars of different strengths and qualities can be made by mixing cements and limes as a binder. There are various additives that, under certain conditions, are added to mortars. These are quick-setting, slow-setting, antifreezes, hardeners, seaters, binding compounds, strain preventatives, waterproofers, curing compounds, air entertainment materials, plasticity improvers, colors, and many other materials in plastics, powders, liquids, and crystals. Straight limes and cements, separately or together in different proportions, are all the binders needed to make the best mortars. Water, free from impurities, and good graded aggregates are needed to make mortar. Color can be added, if required. When making a colored mortar mix, color is more evenly blended when it is mixed with the binder before the binder is added to the aggregate.

There are several other brand binders, cements, and ready-mixed materials on the market. Their mixing methods are generally on their containers. Some are Ideal, Atlas, Brexment, Brickcreet, Lehigh, Hytest, Wifco, Brickset, and Keene's. A percentage of lime or cement can be added to these if desired. Water is required to start the chemical action when making mortar.

AGGREGATES

Sand is widely scattered on the earth. Its loose unconsolidated grains and fragments were once contained by silica, calcium carbonate, clay, iron oxide, calcium sulfate, or other bonding materials. Sand, generally hexagonal, is a good aggregate. It is vitreouslike and either dull, translucent, or opaque. It is tinted according to its metallic oxide content. It is classified as to size, hardness, sharpness, weight, porosity, and resistance to the elements. It is graded as to size when used in mortars. Various sizes are incorporated for the purpose of lessening the voids in mortar.

Sand is a natural aggregate. There are other kinds of aggregates within the natural aggregate group.

Natural Aggregates

Natural aggregates include sand, gravel, crushed rock, stone, granite, marble, shells, slate, pumice, shale, scoria, etc. These have to be crushed and graded before use.

Artificial Aggregates

Artificial aggregates include broken brick, broken and crushed clay products, ashes, glass, artificial slag, iron filings, slag, pottery, clinkers, etc. These have to be graded before use.

Fibrous Aggregates

Fibrous aggregates include wood, straw and hay, fiber, reeds, coir, cork, hair, shavings, sawdust, taw, etc. These have to be processed and graded before use. They are seldom used in masonry mortars.

Graded Aggregates

The particles of an aggregate for masonry mortar should vary in size. There are few voids in masonry mortar if graded aggregate is used. This aggregate should let 100 percent of its particles pass through a number 4 screen (a screen that has ¼" holes). Ninety-five percent of the particles, should pass through a ⅛" opening, 65 percent should pass through a 1/16" opening, 30 percent should pass through a 1/32" opening, 20 percent should pass through a 1/50" opening, and 10 percent should pass through a 1/100" opening.

Variations in these percentages may make the graded aggregate either coarse or fine. Should the aggregate be on the coarse side, more binder is needed in the mortar. Should the aggregate be on the fine side, less binder is needed. Should the aggregate be on the coarse side, less water is required. Should the aggregate be on the fine side, more water is needed in the mix. When the mortar is on the coarse side or the fine side, its strength is not the best. A well-graded aggregate has good plasticity and tensile strength. It will not dry out or set up too rapidly as do coarse or fine mortar mixes when used.

All aggregates used in masonry mortars should be free from harmful ingredients such as dust, silt, clay, vegetable matter, salt, sulfur, alkali, etc. Practically all natural aggregates are found with other materials of varying quality and size. If an aggregate is not taken from a sand pit of known quality, it should be tested for impurities.

To test an aggregate for dust, silt, and clay content, the *settling method* is used. Place a 2" depth of the aggregate in a pint jar. Fill the jar with drinkable water (water free from impurities). Shake the jar and let it stand until the water above the aggregate is clear. If the layer of silt, dust, or clay is more than ⅛" in depth, the aggregate should not be used in a mortar mix until it is freed of the unnecessary material. It should be washed free of most of the dust, silt, clay, etc.

By adding to this aggregate a spoonful of common lye before washing, the aggregate can be tested for undesirable matter. Should the water above the aggregate be darker than apple cider vinegar, the aggregate should be washed.

Aggregate containing alkali of sulfur is easily detected by sight and smell. This aggregate should be washed free of these impurities and aired well for several days. An aggregate requires washing to free it of salt.

The best aggregate to use is aggregate that is free from impurities which weaken a mortar. A mortar is no stronger than the aggregate used. A well-graded and impurity-free aggregate should be used.

COLOR

Units of masonry construction have their own natural color. This color is put there by the materials of which the units are made. Mortars reflect the colors of the materials of which they are made. Actual colors can be put in mortars and the mixes of which artificial construction units are made.

The best colors to use are *mineral* colors. They will not fade or leach off. These iron oxide colors stay put. Different color manufacturers have special names for the various colors they make. There are maroons, bright and dull reds, medium reds, brownish reds, vermilions, browns, Italian reds, metallic browns, rich browns, light browns, chocolates, jet blacks, blue blacks, supreme blacks, metallic blacks, and mineral blacks. There are yellows, blues, greens, purples, and combinations of these colors.

Color powders come in small or large quantities. A trial mix of the desired color gives the quantity of color needed for the certain depth of colored mortar required. Not over 10 percent of color to the binder should be used in a mortar mix. This quantity varies from 3 to 9 pounds of color per sack of portland cement used. Some nonmineral colors require more color and leave a weakened mortar mix. Care should be taken to evenly mix the color well in the batch, or the set mortar joint will have a mottled appearance. Jointing of colored mortar joints should be done after the colored mortar joint is in its semiset stage, or the color will run.

Care should be taken when washing down masonwork that has colored joints. Color can, if the mortar is not cured, bleed out and color the units of masonwork.

WATER

Good, clean water should be used when making all mortar mixes. It should be free from acids, alkalies, oils, sulfates, and any impurities that weaken or harm the mortar. Sea water should not be used.

Salt has the property of attracting moisture from the air. Moisture causes an efflorescente to appear on the surface of masonwork, whether the mortar is set or fully cured, especially when salt is in the mortar. Salty water should never be used if the set mortar comes in contact with chemicals or ammonia. The salt water mix becomes valueless if this happens.

Never use undrinkable water. The quantity of water to use in a mix should be no more than is necessary to wet every part of the mix. An excess amount of water impairs the quality of the mix.

When an excess amount of water is used, as when making thin mortar, more binder is needed to offset the weakening of the mortar.

The purpose of water in a mortar mix is to saturate the binder with enough water to start the binder's chemical action, giving the mortar the proper consistency to make it easy to use. Mortar should be used before it starts its initial set. Some mortars can be tempered with water when they start to take up.

MORTAR MIXES

A good, strong mortar ingredient proportionment, is:

- 1 part portland cement.
- 3 parts good graded sand.
- Water as needed.
- Color to 10 percent can be added to the binder, if required.
- 10 percent hydrated lime added to the cement binder helps retard water after setting.

Another good mortar proportionment is as follows:

- 1 part masonry cement.
- 2½ parts good graded sand.
- Water.
- Color, if necessary.
- 10 percent to 15 percent lime can be added if the mortar is to be used below the water. Lime is always mixed with the binder.

A good cement lime mortar is proportioned as follows:

- 1 part portland cement.
- 25 to 50 percent part of lime or lime putty.
- 4 to 6 parts of good graded sand.
- Water. Color, if required.

A good proportionment of lime and portland cement mortar is as follows:

- 1 part portland cement.
- 2 parts lime or lime putty.
- 5 parts graded sand.
- Water. Color, if required.
- A good mortar mix for freestanding and nonload-bearing partitions.

A good mix for flue or chimney work is as follows:

- 1 part portland cement.
- ¾ part hydrated lime or putty.

●6 parts good graded sand.
●Water. Color, if required.

Cement mortar is much harder to handle with a trowel than is lime mortar.

METHODS

Mortar is used as joint material when laying or setting masonry units. Mortar not only holds the units apart, but also fills in the voids between the units. Good plastic mortar, when set in joints, holds units together and in place.

Mortar is composed of aggregate and binder. There are several kinds of aggregates and binders.

Hand or power tools are needed to make mortar. Dried mortar stuck to the tools may come off and get in the new mortar. When this happens, the dried mortar does not work well with a trowel. This dried mortar does not take up moisture. When placed in a joint, it causes a "dry-out." This dry-out is subject to the elements and, in time, ruins the joint.

There are two kinds of methods used when making mortar on a job. The *hand method* is generally used, although the *machine method* may also be tried.

Hand Method

The items required for this method are a mortar box, hoe, shovel, screen, some pails, a barrel, and a water hose. A measuring box or pail is a necessity.

The screen is necessary to screen the aggregate, even though it is graded aggregate. This should be a screen having four holes to the square inch. The binder should be on hand. Put the water in the barrel.

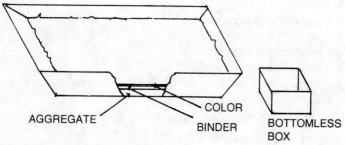

AGGREGATE COLOR BINDER BOTTOMLESS BOX

Fig. 3-4. The hand method of making mortar.

Put the screened sand in the measuring box which is in the mortar box. Keep count of the measures of sand. Level off the aggregate. The binder should be measured for accuracy. Put the binder on top and level it off. Chop the aggregate and binder with the hoe. Chop them from one end of the box to the other and back again until every particle of the aggregate is covered with binder.

When color is used, it should be thoroughly mixed with the binder before the binder is put on the aggregate. Water is slowly added a little at a time and mixed (Fig. 3-4).

Machine Method

When machine mixing is done, measuring pails, shovels, aggregate, binders, and water are needed. The machine mixer should be level. Position a wheelbarrow to catch the mortar from the machine mixer. Some water is put in the operating mixer first to keep the mix from sticking to the machine mixer.

The aggregate is added next. The binder is now added along with more water. Water is used as necessary to make a good mix, either (thick) or thin mortar as desired. Whether hand mixing or machine mixing, all tools used should be clean.

Chapter 4

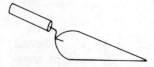

Masonry Units

There are various masonry units used in masonwork. They serve the purpose for which they are intended. Note the stone units in Fig. 4-1 and the brick units in Figs. 4-2 through 4-4.

MAKING BLOCKS

There are many kinds of block manufactured. They fall into two general classifications: heavy- and lightweight. They are made by hand or by machinery.

Today blocks are made mostly by machinery in elaborate manufacturing plants. They are made with a mixture composed of a binder and an aggregate. Additives such as color (chloride, color combinations, etc.) and calcium chloride are added percentagewise.

Block construction units are made with a mix composed of cement and percentages of graded aggregate. Color can be added to make colored block. Water is used sparingly to start the chemical action.

Regular, high-strength, or air-entrained (5 to 8 percent) portland cement is generally used as a binder. Natural aggregate produces a stronger load-bearing and a more durable block.

A hollow-cored block (8″ × 8″ × 16″) weighs from 40 to 50 pounds. The same size block made with pumice, expanded shale, or lightweight artificial aggregate, weighs from 22 to 30 pounds. The lightweight block has the added advantage of nailability, increased insulation, and sound absorption. Properly made, the blocks have low capillary moisture absorption.

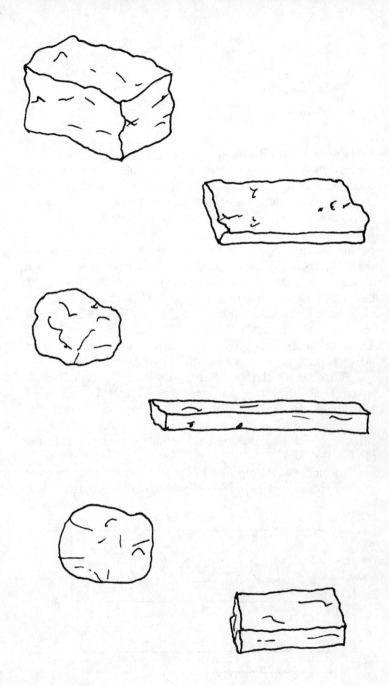

Fig. 4-1. Stone units.

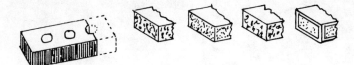

Fig. 4-2. Textured bricks.

Forms used in block making are of different sizes and shapes. Various effects can be had from different molds. Block sizes are becoming standardized in that block sizes conform to modular sizes rather than the standard sizes of brick. (Brick sizes are also drifting to the modular sizes.) The modular sizes will, with the mortar joint, fit into modular construction. The units with the mortar joints can readily fit the heights and lengths, breadths, and the different depths of modular construction. Generally, in modern construction 4′ lengths are considered a unit measure. Any block units that fill this space with the joint are modular. A 2⅝″, 3⅝″, 7⅝″, 15⅝″ size block length is modular. In height a 2¼″, 3⅝″, 7⅝″ size block is modular. In depth a 3⅝″, 5⅝″, 7⅝″, 11⅝″ size block is modular. These are the sizes most generally made in regular block. There are other specially made blocks that fit in modular buildings.

Block is cast solid (like artificial stone) or hollow celled (cored) as structural tile (averaging approximately 45 percent core.) Celled blocks have various shell wall thicknesses. They can be cast in pairs that lock or lap when laid in a wall. Generally, they are cast as a single unit with two or three cells running verticaly through the block.

In a block mix the proportion of binder (cement) and aggregate may vary because of the shape and grading of the aggregates used,

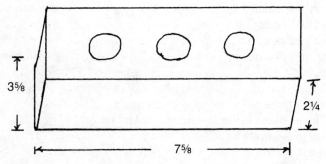

Fig. 4-3. An 8″ whole brick.

40

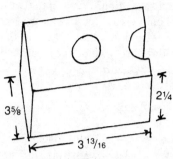
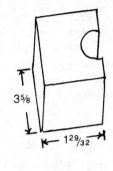

Fig. 4-4. More brick examples.

water quantity, additives, and mixing facilities. Also, the mix varies as to the purpose for which the block is used. Some blocks require load-bearing strength, some need acoustical qualities, and some require fire resistance or water absorption resistance. All blocks made should be required to resist the elements and pass the required crushing test.

A good mix to use when casting block by hand is 1 part portland cement, 3 parts graded sand, 2 parts of ⅜″ coarse aggregate, and just enough water sprayed on when mixing. The water content should be such that a handful of the mix when squeezed does not wet the hand. Another way to test the moisture content is to run a trowel over the mix. If the trowel brings a shine to the mix, and the mix just shows dampness, the water content is right.

This mix is then rammed into the mold or form. The form should be almost immediately lifted from the green block. The mix and the green block should be protected from frost. The uncured or green block should be kept damp for several days and not allowed to dry out for 11 days.

In the manufacture of block in quantity, little or no water is added to the mix. The machinery rams and vibrates the mix in forms under pressure. The green blocks are then saturated with live steam under pressure to start the chemical action of the binder and unite the blocks into a solid and sound construction unit. This method of wetting in the curing room will take from 24 to 40 hours, after which the blocks are placed in the air to continue curing.

In the steam curing room the pressure supplied is from 5 to 10 pounds of steam at a constant of 100 degrees Fahrenheit. The blocks need not remain in the pressured steam room as long as in the unpressured room.

Some block making plants saturate the block before steaming. Some steam them before and after saturation. Some spray with water; others do not. Some plants only steam the blocks. Blocks made by these plants are, when air-dried 36 hours, equal in the strength of air-dried blocks that are three weeks old. The steam and saturation curing process enhances crystallization between the binder and water to solidify the mix into a usable unit.

Lightweight Aggregates

Lightweight aggregates are graded for making lightweight block. They are graded approximately as follows. The total amount of pumice used, or 100 percent, should go through a ⅜″ screen. Of this 100 percent, 82 percent of the pumice should go through a ¼″ screen. Of this 82 percent, 65 percent of the pumice should go through a ⅛″ screen. Of this 65 percent, 65 percent of the pumice should go through a 1/16″ screen. Of this 65 percent, 45 percent of the pumice should go through a 1/3″ screen. Of this 45 percent, 35 percent of the pumice should go through a 1/50″ screen. Of this 35 percent, 25 percent of the pumice should go through a 1/100″ screen. See Table 4-1. The more sand used, the heavier and stronger the block is.

Light-celled aggregate floats in a mix if it is not sprayed with water before using. More water is used when making block with such aggregate. Mixing formulas may be varied to give crushing strength if desired. The less water used shortens the initial set and curing time for a crushing test.

Color and additives may be added to any mix to give the block color and quality. Block can be made in any shape from screen block to solid block, from lightweight to heavyweight, and from small to large sizes. Block is used in load-bearing, nonload-bearing, and ornamental masonry construction.

Block can be cast in various face finished as to shape and texture. Using various size aggregates (colored or natural), blocks

Table 4-1. Mixing Formulas for Pumice Block.

Cement	Pumice		Fine Sand
1 part	10	parts	0 part
1 part	8	part	0 part
1 part	6	parts	½ part
1 part	4	parts	1 part
1 part	2	parts	2½ parts

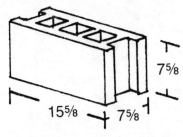

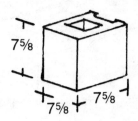

Fig. 4-5. A common block.　　　　Fig. 4-6. An 8" half block.

can show stonelike qualities as to texture. The green-black copper slag (to represent marble) can be an additive to the aggregate. The marble chips with a minimum of mica spar in a buff-colored mix can represent sandstone. Any textured and colored block can be made to represent natural stone.

Sizes and Shapes

Block can be cut, wrought by hand or machinery, and handled as a unit of construction. The various blocks used as construction units can be seen in Figs. 4-5 through 4-23.

The *double square end* block can be cut in half to make two square end blocks. Its use is the same as the *double bullnose* block (Figs. 4-24 through 4-32).

Sash blocks are used next to openings, windows, doors, etc. (See Figs. 4-33 through 4-92).

BONDS

It is pertinent to lap or reinforce all masonry units of construction. This is necessary when masonwork is expected to last a long time.

The different bonds of blockwork not only are pleasing to the eye, but are strongly laid and will withstand lateral and horizontal

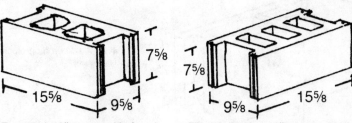

Fig. 4-7. A 10" common block.　　　Fig. 4-8. Another 10" common block.

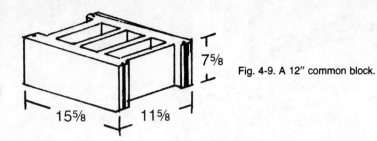

Fig. 4-9. A 12" common block.

7⅝

15⅝   11⅝

pressure. The most common bond is the running *half lap* bond (Fig. 4-93).

The mixed pattern similar to the *ashlar stone* bond is a bond containing various sizes of block. This is a bond that, when laid up in coarse laying, can be reinforced with re-bar or dur-o-wall. It can,

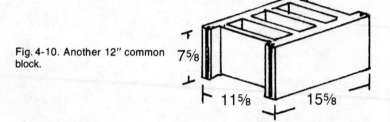

Fig. 4-10. Another 12" common block.

7⅝

11⅝   15⅝

however, be reinforced perpendicularly. This bond is shown in Fig. 4-93, along with the running block bond.

The stacked bond is used in many places, especially in larger buildings. It has to have its joints perpendicular and horizontal. This bond needs reinforcing, or it comes apart. Dur-o-wall reinforcing is generally used (Fig. 4-94).

Another bond is laid with whole blocks. This bond can be reinforced as 16" courses are laid.

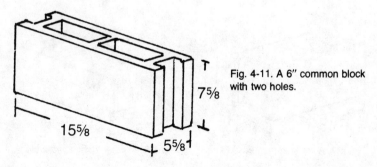

Fig. 4-11. A 6" common block with two holes.

7⅝

15⅝   5⅝

44

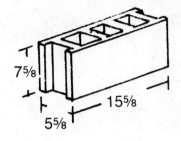

Fig. 4-12. A 6″ common block with three holes.

7⅝

5⅝

15⅝

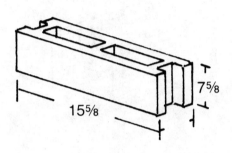

15⅝

7⅝

Fig. 4-13. A 4″ comon block.

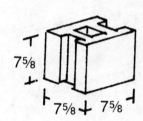

7⅝

Fig. 4-14. An 8″ common half block.

7⅝   7⅝

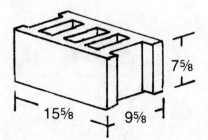

15⅝   9⅝

7⅝

Fig. 4-15. A 10″ common block.

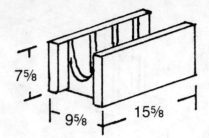

Fig. 4-16. A 10″ bond beam.

7⅝

9⅝   15⅝

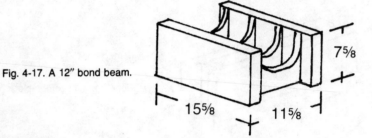

Fig. 4-17. A 12″ bond beam.

7⅝

15⅝   11⅝

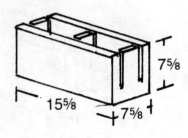

7⅝

Fig. 4-18. An 8″ bond beam.

15⅝   7⅝

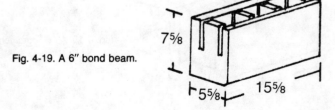

7⅝

Fig. 4-19. A 6″ bond beam.

5⅝   15⅝

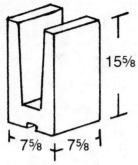

Fig. 4-20. An 8" lintel.

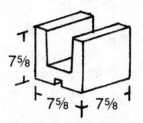

Fig. 4-21. Another 8" lintel.

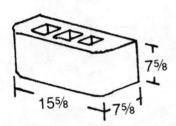

Fig. 4-22. An 8" double bullnose block.

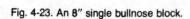

Fig. 4-23. An 8" single bullnose block.

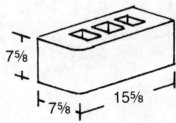

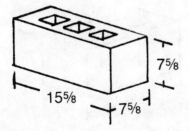

Fig. 4-24. An 8" double square end block.

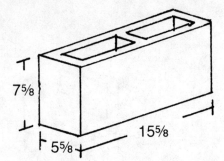

Fig. 4-25. A 6" double square end block.

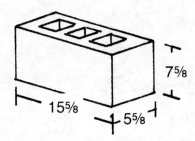

Fig. 4-26. Another 6" double square end block.

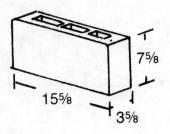

Fig. 4-27. A 4" double square end block.

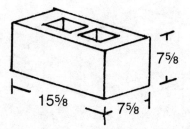

Fig. 4-28. An 8" double square end block with two holes.

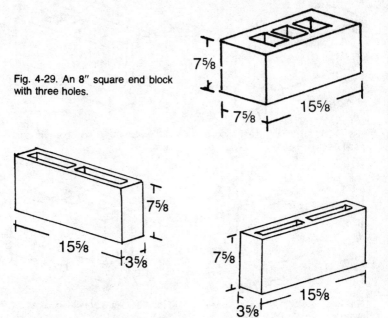

Fig. 4-29. An 8″ square end block with three holes.

Fig. 4-30. Four-inch double square end blocks.

Fig. 4-31. A 4″ half high block.

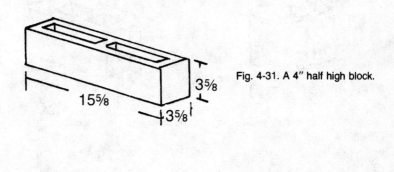

Fig. 4-32. Another 4″ half high block.

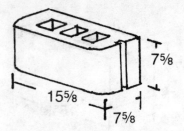

Fig. 4-33. An 8″ double bullnose sash block.

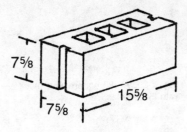

Fig. 4-34. An 8″ double end sash block.

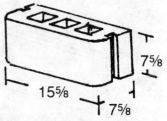

Fig. 4-35. An 8″ double bullnose two-end sash block.

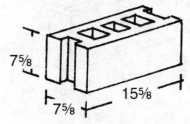

Fig. 4-36. An 8″ one-end common sash block.

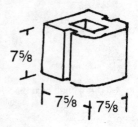

Fig. 4-37. An 8″ bullnose half sash block.

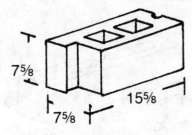

Fig. 4-38. An 8″ joist block.

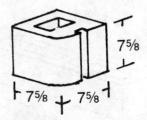

Fig. 4-39. Another 8″ bullnose half sash block.

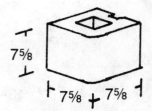

Fig. 4-40. A bullnose half sash block.

50

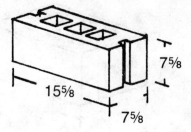

Fig. 4-41. An 8" double square end sash block.

7⅝

15⅝

7⅝

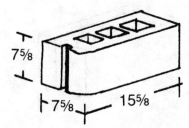

7⅝

7⅝   15⅝

Fig. 4-42. An 8" bullnose sash block.

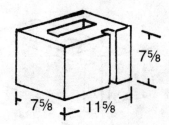

7⅝

Fig. 4-43. A 12" square end sash half block.

7⅝   11⅝

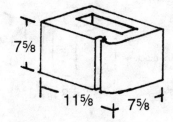

7⅝

11⅝   7⅝

Fig. 4-44. A 12" bullnose sash half block.

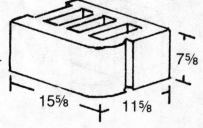

Fig. 4-45. A 12" bullnose sash block.

7⅝

15⅝   11⅝

51

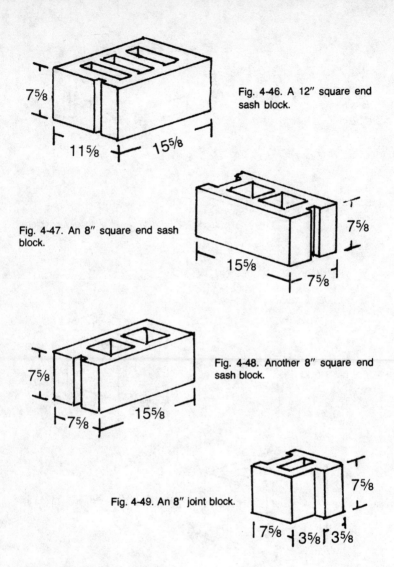

Fig. 4-46. A 12″ square end sash block.

Fig. 4-47. An 8″ square end sash block.

Fig. 4-48. Another 8″ square end sash block.

Fig. 4-49. An 8″ joint block.

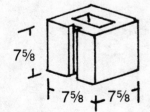

Fig. 4-50. An 8″ half square end sash block.

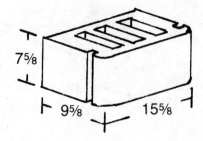

Fig. 4-51. A 10" double bullnose sash block.

7⅝
9⅝
15⅝

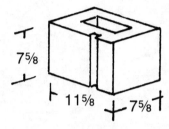

7⅝
11⅝   7⅝

Fig. 4-52. A 12" half sash block.

Fig. 4-53. A standard flat block.

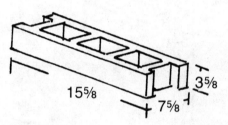

15⅝
3⅝
7⅝

Fig. 4-54. A standard flat sash block.

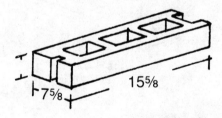

15⅝
7⅝

Fig. 4-55. A bullnose flat sash block.

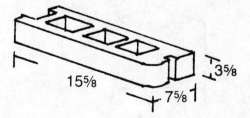

15⅝
3⅝
7⅝

53

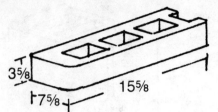

Fig. 4-56. A bullnose flat block.

$3\frac{5}{8}$

$15\frac{5}{8}$

$7\frac{5}{8}$

Fig. 4-57. A half flat block.

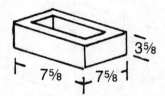

$3\frac{5}{8}$

$7\frac{5}{8}$ $7\frac{5}{8}$

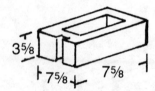

Fig. 4-58. A half flat sash block.

$3\frac{5}{8}$

$7\frac{5}{8}$ $7\frac{5}{8}$

$9\frac{5}{8}$

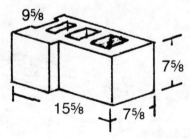

Fig. 4-59. A 10″ offset block.

$7\frac{5}{8}$

$15\frac{5}{8}$ $7\frac{5}{8}$

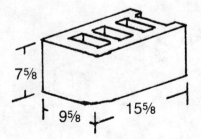

Fig. 4-60. A 10″ bullnose block.

$7\frac{5}{8}$

$9\frac{5}{8}$ $15\frac{5}{8}$

54

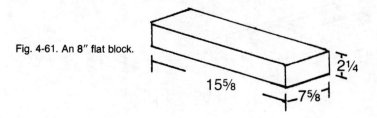

Fig. 4-61. An 8″ flat block.

15⅝

2¼

7⅝

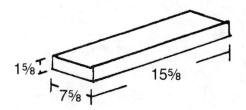

Fig. 4-62. An 8″ flagstone block.

1⅝

15⅝

7⅝

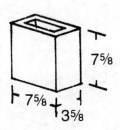

Fig. 4-63. A half partition block.

7⅝

7⅝

3⅝

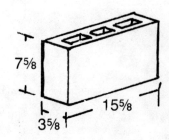

Fig. 4-64. A 4″ partition block.

7⅝

15⅝

3⅝

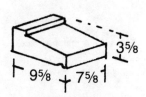

Fig. 4-65. A sill block.

3⅝

9⅝

7⅝

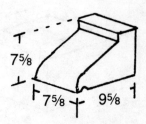

Fig. 4-66. Another sill block.

7⅝

7⅝

9⅝

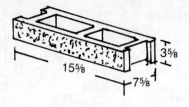
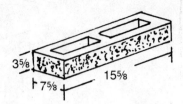

Fig. 4-67. Textured blocks.

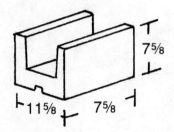

Fig. 4-68. A beam block.

Fig. 4-69. Another beam block.

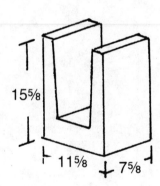

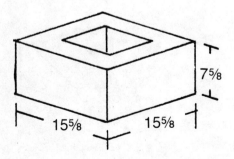

Fig. 4-70. A chimney block.

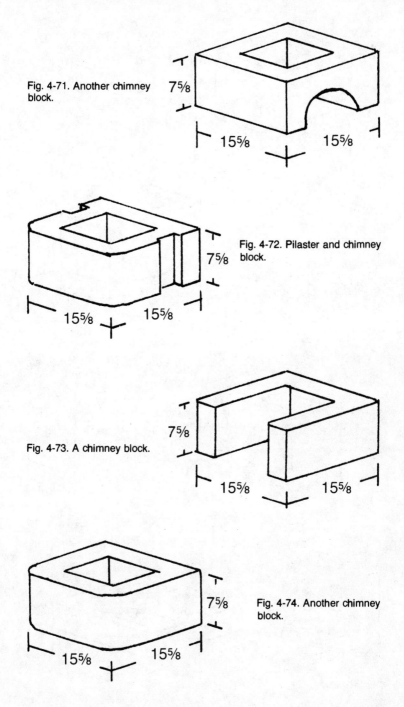

Fig. 4-71. Another chimney block.

7⅝

15⅝   15⅝

Fig. 4-72. Pilaster and chimney block.

7⅝

15⅝   15⅝

Fig. 4-73. A chimney block.

7⅝

15⅝   15⅝

Fig. 4-74. Another chimney block.

7⅝

15⅝   15⅝

57

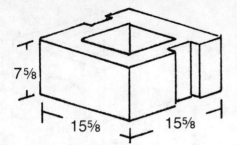

Fig. 4-75. A pilaster block.

$7\,5/8$

$15\,5/8$ $15\,5/8$

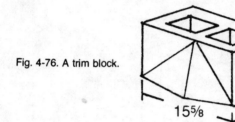

Fig. 4-76. A trim block.

$7\,5/8$

$15\,5/8$ $7\,5/8$

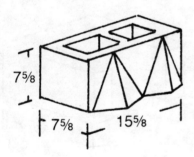

Fig. 4-77. Another trim block.

$7\,5/8$

$7\,5/8$ $15\,5/8$

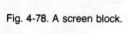

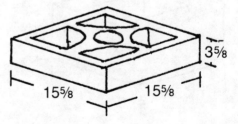

Fig. 4-78. A screen block.

$3\,5/8$

$15\,5/8$ $15\,5/8$

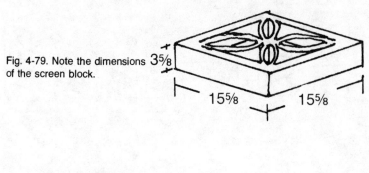

Fig. 4-79. Note the dimensions of the screen block. $3\frac{5}{8}$ $15\frac{5}{8}$ $15\frac{5}{8}$

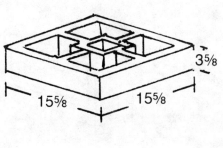

Fig. 4-80. Note the holes in the screen block. $3\frac{5}{8}$ $15\frac{5}{8}$ $15\frac{5}{8}$

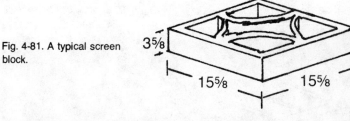

Fig. 4-81. A typical screen block. $3\frac{5}{8}$ $15\frac{5}{8}$ $15\frac{5}{8}$

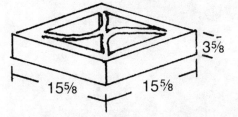

Fig. 4-82. Another screen block. $3\frac{5}{8}$ $15\frac{5}{8}$ $15\frac{5}{8}$

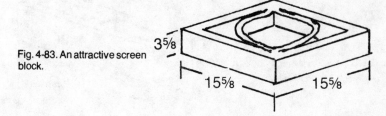

Fig. 4-83. An attractive screen block. $3\frac{5}{8}$ $15\frac{5}{8}$ $15\frac{5}{8}$

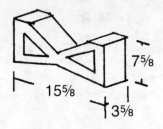

Fig. 4-84. A 4″ screen block.

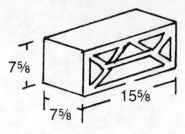

Fig. 4-85. An 8″ screen block.

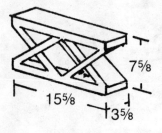

Fig. 4-86. Another 4″ screen block.

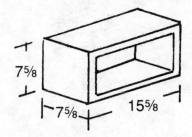

Fig. 4-87. Another 8″ screen block.

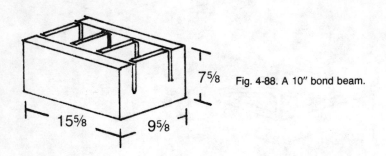

Fig. 4-88. A 10″ bond beam.

Fig. 4-89. Another 4″ screen block.

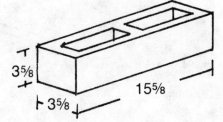

60

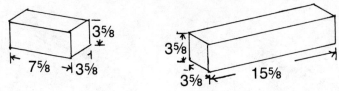

Fig. 4-90. Four-inch jumbo blocks.

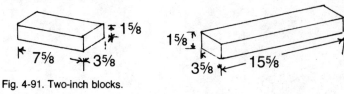

Fig. 4-91. Two-inch blocks.

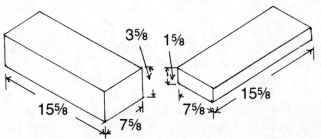

Fig. 4-92. Four-inch and 2″ solid flats.

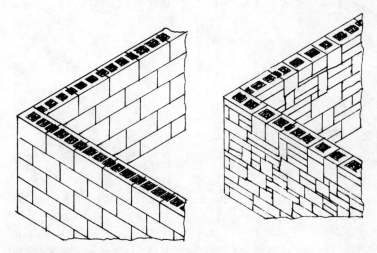

Fig. 4-93. Block bonds.

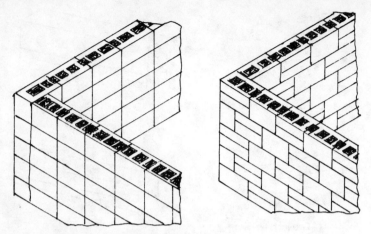

Fig. 4-94. Dur-o-wall reinforcing is used in this bond.

The bond laid in 8″ breadth courses can easily be reinforced. Every other whole block is laid double and separated by split block. The blocks are half lapped. The block bond in Fig. 4-95 consists of one course of whole blocks alternated with two courses of split block.

The block bond containing longitudinal whole block, along with perpendicular laid block encircling a half block, is a pleasing bond to the eye. I have never reinforced this bond (Fig. 4-96).

The method of block layout with respect to 8″ and 6″ block is shown in Fig. 4-97. The names of the different parts of construction,

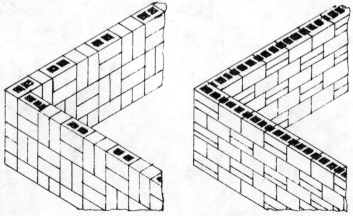

Fig. 4-95. This block bond has one course of whole blocks alternated with two courses of split block.

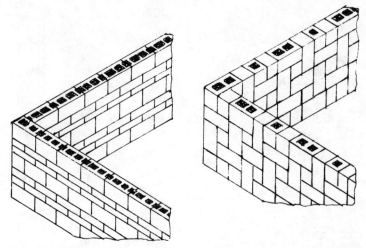

Fig. 4-96. An attractive block bond.

as well as where they are located, are shown in the drawing of a sectional cut of a wall (Fig. 4-98).

The method of handling and placing mortar is unique with each bricklayer. One good method is shown in Fig. 4-99. The line is also a necessary tool.

The method of seating the mortar to the trowel is easy. Just take a trowelful of mortar, lower the trowel fast, and quickly stop its downward motion. This seats the mortar to the trowel so it will not slip off until it is used. The trowel holds this mortar when it is turned upside down.

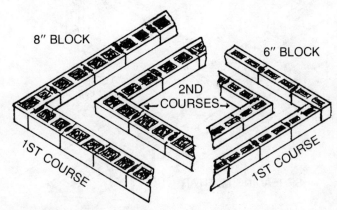

8″ BLOCK

6″ BLOCK

2ND
COURSES

1ST COURSE

1ST COURSE

Fig. 4-97. Block corners.

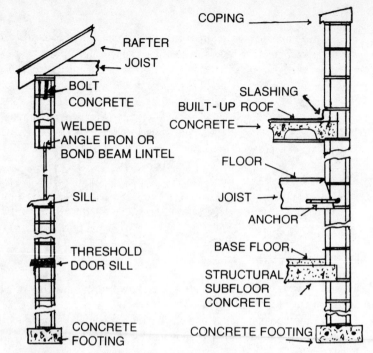

Fig. 4-98. Sectional cut of a wall.

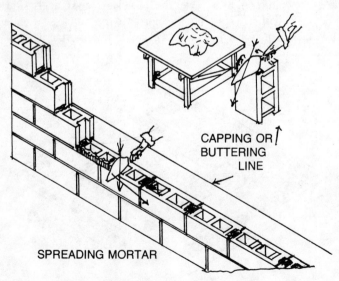

Fig. 4-99. A method for handling mortar.

64

TILE

Tile is manufactured either hard with dense texture, or porous and lighter than other tile units of equal size and shape. The porous tiles have more sound absorbing ability and better insulation quality. Porous tiles are made like various hard clay tile units, except that straw, sawdust, or other combustible aggregate materials are added to the clay mix. This added aggregate is burned up, leaving the tile porous. Its insulation and acoustical qualities are enhanced as well as its weight. Porous tile is made with at least 1″ thick walls (shells) and ¾″ thick webs to insure strength.

Clay tile units are made with single or double shells. When they are set in masonwork, they are laid with their shells horizontal or vertical, whichever way strength is needed. Tile may be enameled or glazed. Tile can be colored by special treatment or by degrees of burning.

Structural Tile

Structural tile is made in many shapes, styles, and sizes, and of different materials for interior and exterior masonwork. Tiles are used in walls and in backup masonwork—to cover beams, girders, and columns—and as finished masonwork. Tile types include wall tile, furring tile, partition tile, book tile, flue tile, drain tile, angle tile, key tile, interior tile, arch tile, decorative tile, and exterior tile. Other tiles are load-bearing tile; glazed tile; smooth, scored, or tapestry tile; sanitary tile; acidproof tile; fireproof tile; and acoustic tile. There are various structural tiles used in masonwork, both inside and outside (Fig. 4-100).

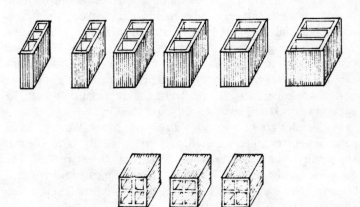

Fig. 4-100. Structural tiles.

Clay tile is burned various times for different tile. After burning, clay tile can be glazed by coloring it (the biscuit) with liquid made from pulverized clay, flint, or feldspar fluxed by heat. After coating, the biscuit is placed in the burning kiln and heated to vitrify and fuse the liquid glaze to the clay tile. In these glazes or surface finishes, the glazes and enamels, colored or clear, are proportioned and fired to produce the desired texture such as mat, dull, high, or low glaze. Any intermediate texture or finish can also be had. Many color effects can be produced, especially by different firings. Different temperatures have their effects on the tile when fired. By using various metallic oxides which color or stain the glaze while it is vitrifying many kinds of tile finishes can be produced. The color of clay tile never fades once it is a part of the finished tile.

Clay tile units that can be fired to complete vitrification in one burning are the hardest and cannot be scratched. Some clay tile units cannot be fired; thus, they are classed as semivitreous tile. These tiles are the best outside tiles. They are nonabsorbent, durable, and waterproof. Another clay tile is made with an abrasive which is added to the basic clay mix.

Floor and Wall Tile

Floor and wall tiles are thin, platelike units. They are of various sizes, thicknesses, shapes, colors, finishes, and of many different qualities. They are used as coverings for counter tops, and as floor tile, wall tile, and decorative panels. They vary in quality and appearance and should be stored previous to laying. There are inside and outside tiles.

The decorative tile is classified as paving, bright (or mat-glazed or enameled), unglazed, vitreous, unglazed or glazed vitreous, plastic, inlaid, quarry, and mosaic. Further, within these classifications are groupings as to hardness, size, color, finish, and qualities respecting grade and durability. Tile can be made in any contour necessary to fit any required place. Small sizes of tile are mounted on paper in any design. These sheets are laid as a unit. The backing is removed, and the joints are filled.

Clay tiles are made in many sizes and shapes. When glued on paper sheets (or other material), they are called tile mats. The mat sizes range from small areas to as much as 9 square feet. The thickness varies from ¼" to 1". Various designs and decorative pictures, due to the various tile placement on these mats, are formed (Fig. 4-101).

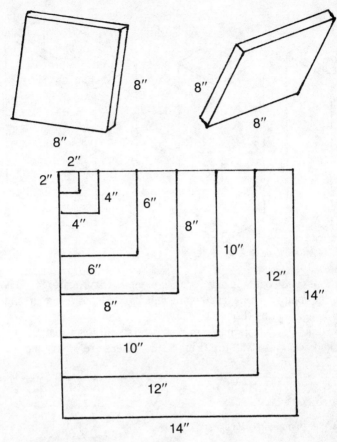

Fig. 4-101. Floor and wall tiles.

Flue Tile

Flue tile is made round, square, rectangular with rounded corners, oval, and in various sizes and lengths. Many flue tile units are glazed. This glazed tile outlasts the unglazed tile (Fig. 4-102).

Drain Tile

Drain tile is porous, solid, or perforated. It can be belled and is made in various sizes. The drain tiles are made either round, square, or oval, with straight or bent lengths. The sizes, shell thicknesses, and lengths vary (Fig. 4-103).

Coping tile is made in different size widths, shapes, and thicknesses. There are clay and concrete coping tiles (Fig. 4-104).

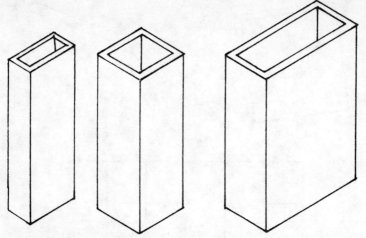

Fig. 4-102. Flue liner tiles.

The solid coping is made with cement and aggregate. Stone is used as coping. There are coping blocks made of a binder and various aggregates (Fig. 4-105).

Some clay tile is made in a structural tile form that can be knocked apart into wall tile (Fig. 4-106). There are other kinds of

Fig. 4-103. Perforated grain tile and drain tile.

tile made for certain uses in masonry construction (Figs. 4-107 through 4-109).

Roofing Tile

Roofing tiles are made in many sizes and shapes. They shed water when installed as roofing or as a decoration. Some roofing

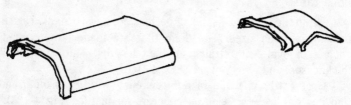

Fig. 4-104. Coping tile and a coping tile corner.

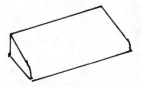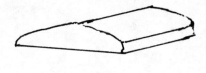

Fig. 4-105. Solid coping tiles.

tiles have projections on the underside that catch the roof cleats (wood or metal cleats), or they have projections that catch on laid tile. Then there are those that have holes that secure them when nailed or wired.

Stone Tile

Stone can be laid as tiles on floors and walls. Fabricated stone tile is also used as tile. Here the thickness varies. Slate, when split correctly, makes a good tile on floors, walls, roofs, and patios.

The fabricated tile is made from a mix having different percentages as well as aggregates. Some are even partition blocks

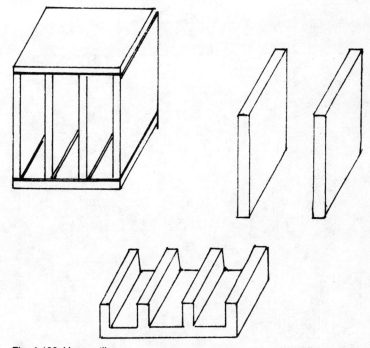

Fig. 4-106. Veneer tiles.

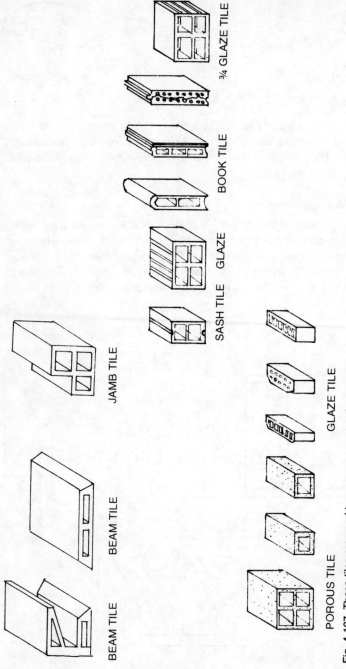

BEAM TILE

BEAM TILE

JAMB TILE

SASH TILE   GLAZE

BOOK TILE

¾ GLAZE TILE

POROUS TILE   GLAZE TILE

Fig. 4-107. These tiles are used in masonry construction.

70

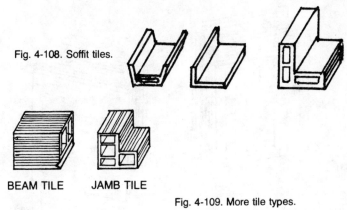

Fig. 4-108. Soffit tiles.

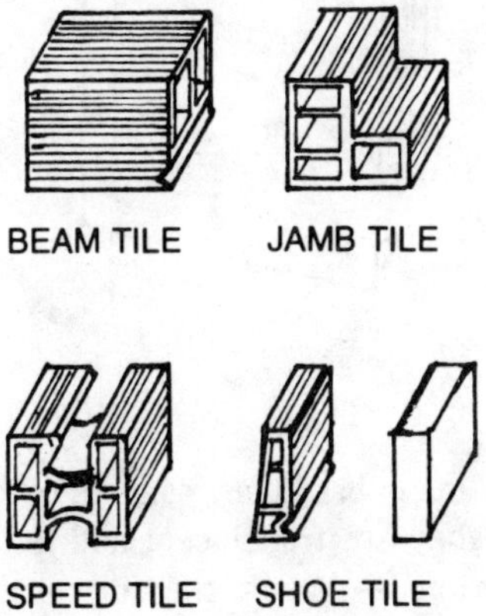

BEAM TILE   JAMB TILE

Fig. 4-109. More tile types.

SPEED TILE   SHOE TILE

made from gypsum. Other tiles are made from asphalt, plastic, and metal. Slab planks and decorative panels are made from cements, plastics, metals, gypsum, magnesia, glass, etc. Many are made dense, textured, smooth, heavy, lightweight, porous, sound and heat-resistant, waterproof, and acid-resistant.

Some tile units and tilelike planks are porous and nailable. Some tiles need various kinds of mortar when used (Fig. 4-110).

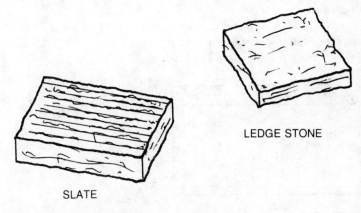

LEDGE STONE

SLATE

Fig. 4-110. Slate and ledge stone can be laid as tile.

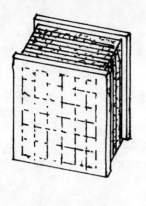 

Fig. 4-111. Glass blocks.

GLASS BLOCKS

Glass blocks are called tile by many brickmasons as they are partially evacuated; being hollow, they are translucent and let in light. Some glass blocks are directional—that is, they direct the light which passes through them. They are designated on their top so that the mason may set them directional as required. The light is directed at the ceiling, to either side or down as called for by the builder. Glass block is not easy handling block. The blocks are generally slick and can slip out of your hands easily. They are heavy enough to settle mortar and reduce the size of the bed joints if the mortar is not heavy in density. Therefore, the mortarman should not make sloppy mortar when glass blocks are to be laid or set. Care in the handling of glass blocks by the workmen lessens the possibility of breakage.

Glass block as a unit of construction cannot be surpassed by any other unit of construction when set right. Figures 4-111 through 4-114 illustrate some glass blocks.

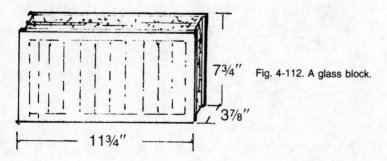

7¾" Fig. 4-112. A glass block.

3⅞"

11¾"

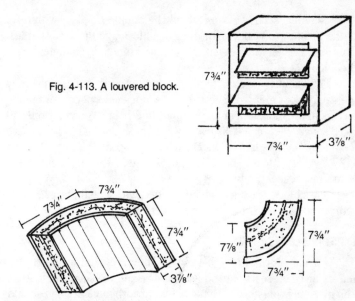

Fig. 4-113. A louvered block.

Fig. 4-114. More glass blocks.

Joints of glass masonry should be finished with a concave jointer when the mortar has reached a firmness in its initial set. Wipe clean the finished work before the mortar is fully set.

Because glass blocks are subject to expansion, panels should be kept within 22 to 26 square feet of the face surface, and not more than 4 to 6′ in length, nor more than about 7′ in breadth, unless expansion joints are used. They should be recessed (Fig. 4-115).

Fig. 4-115. Glass blockwork.

Caulking, such as *oakum*, should be rammed in place within ¾″ of the finished joint between the glass block and the recess sides. This joint is then filled in with mortar and tooled in alignment with the jamb.

The quantity of glass block needed for a job is determined by the square foot method. Because glass block does not absorb moisture from the mortar, a brickmason in an 8-hour day can lay only 40 to 50 square feet of 3⅞″ by 5¾″ by 5¾″ block, or 100 to 125 square feet of 3⅞″ by 7¾″ by 7¾″ block, or 115 to 135 square feet of 3⅞″ by 11¾″ by 11¾″ block. These quantities take 5, 3½, or 2½ cubic feet of mortar, respectively.

Glass block masonry can be cleaned with a cloth, brush, soap, and water. A wire brush is never needed.

GYPSUM BLOCKS

Gypsum blocks are lightweight. They are used for partitions and are also fire-resistant. They should be laid with gypsum mortar. The gypsum is plastered. These blocks absorb moisture and should be laid on other units as a first course (Fig. 4-116).

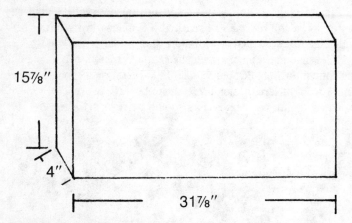

Fig. 4-116. A gypsum block.

MAGNESIA BLOCKS

Magnesia blocks require magnesia mortar. Regular masonry mortar does not hold magnesia blocks, as magnesium does not mix with cement mortar. The method of using this block in a construction job is as follows. Plaster the wall to which the black will adhere with two coats of magnesium mortar. Before this is semidry, the magnesia blocks should be splashed with water on its backside and quickly placed with pressure into the plaster mortar. When held a short time, it seats itself. If this block is not immediately placed when wet, it falls apart. Mortar is tucked in between the blocks after they are laid. These blocks are sound-resistant. They are very porous and do not sustain a load (Fig. 4-117).

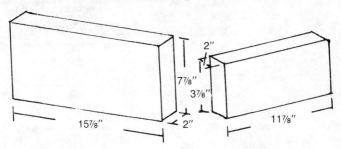

Fig. 4-117. Magnesia block.

TUCKING WORK

Tucking comprises repair, overhaul, and refurbishing masonwork. Generally, in masonwork the units are set or laid, the joints are finished, and the cleaning, pointing-up, and washing are all performed by the mason. There is, however, masonwork that is not finished or old and needs repairing. Such masonwork requires a tucking mason to perform this work.

The mortar used by the tucking mason is generally as near the same mix as possible as that which was used in the masonwork. A different mix may be called for when tucking masonwork. Color may be required, especially on old masonwork.

A good tucking job requires that the joints be raked out properly and cleaned out before the mortar is applied to the well-wetted joints. After any loose, broken, or weathered masonry units are replaced, the joints needing repair should be raked out to a depth of 1″ (Fig. 4-118). Then the joints should be wetted with water to stop suction, before they are filled with mortar. Suction of the masonwork will dry out the mortar before it has a chance to set up. A dried out mortar joint is worthless. Mortar should be placed in the joint in layers. Each layer should be pressed well with a tucking tool until the joint is filled. After the mortar joint takes up, it should be kept damp at least until it is set. It will then cure to its good hard state.

It is easier to fill the joints by using a long narrow tuck-pointer which is slightly smaller in width than the joint. The mortar can be held with either the back of a trowel or a tuck mason's hawk.

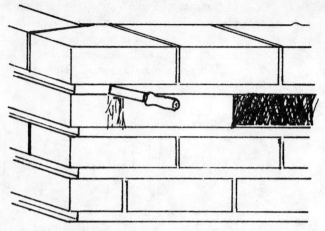

Fig. 4-118. Joint raking and repairing.

On tucking up new masonwork, the mortar should be like the mortar used in the masonwork. On old masonwork, the mortar can be whatever is desired. A good mortar mix is as follows: 1 part portland cement to 2 parts fine graded sand. Color as required to not over 10 percent of the binder may be used. Mix well and add water to start the chemical action. Mix only enough mortar that can be used in one hour. Mortar for brick, block, and tile can have a percentage of lime in the binder, from 10 percent up to 50 percent lime.

On old masonwork, after the necessary units are replaced, the old joints are raked out to a depth of 1″. The joints should be saturated with water to eliminate suction. Mortar is applied in compressed layers until the joints are well-filled. The joints can be tooled when they are semiset. Should the old masonwork be worn to the point that the units are ragged and uneven next to the joints, the following method can be used when tucking. After the joints are raked out, saturated with water, and tucked with mortar, the joints can be struck or tooled with a recessed jointer. Use a straightedge for this tooling, or the tooled joint will be crooked. This tooled joint is then filled and struck off with a trowel. The mortar can be colored mortar which will give the finished joint a contrast to the masonwork.

Masonwork is now cleaned and washed down after the joints are set. Water having 10 percent to 15 percent muriatic acid or some other good wash is used. If an acid wash is used, the masonwork should be washed down with clear water within an hour after the acid wash.

Ladders and swinging scaffolds are generally used by the tucking mason. The regular mason's and builder's scaffolding is used if possible. All tools used by the tucking mason should be clean and free from set mortar.

Chapter 5

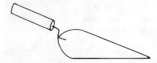

Plastering

Plastering is the application and manipulation of plaster mortars into a protective cover of other construction material, including masonwork. Plaster is often a decorative cover. There are some workmen who call plastering stuccowork; in this book plastering shall be considered as inside work and stucco as outside work. In first-class plastering the variation in trueness of the finish coat of plaster should not exceed 1/16″.

Plastering and stuccoing are done in much the same order. The mortar mix varies somewhat in both. Many plaster mixes do not, after application, withstand moisture, whereas stucco mixes do. Stucco mixes are used on inside work as well as on outside work. Plaster mixes are used only on inside work. Plaster does not withstand the elements as does stucco.

Today plasterwork and stuccowork are done in two-coat jobs. Formerly, the work was done in three or more coats (applications). When machines apply either plaster or stucco, one-coat work is done. A truly first-class job of either plastering or stuccoing is the three-coat method.

Plaster is applied as a cover coat on all kinds of masonwork. It is applied on wood lath, which is spaced approximately ⅜″ apart for a plaster key. When applied on wood lath, the lath has to be sprayed well with water. The lath swells when wet. When the plaster is applied to wet swelled lath, it does not readily crack as the lath dries out and shrinks. This leaves plenty of unpressured space like plaster cleats that the plaster hangs on. When plaster is applied to gypsum board, whether plain, perforated, or reinforced, it does not need its base wetted. When applied to insulation board made with moisture semiproof fiber such as wood, cane, stubble, gypsum, or asbestos, and metal lath or paper interlocked with wire, the base does not have to be sprayed with water.

Wood lath is generally nailed to wood framework. The composition lath is also nailed to wood framework or a wood base. Metal lath and interwoven paper with wire are fastened to woodwork and frames (metal or wood). The metal lath comes in sheets or rolls and is handy as a plaster or stucco base. It can be fastened to hangers, on walls or ceilings, and to any odd-shaped places. Some V-ribbed metal lath or rod-stiffened expanded metal lath can stand alone. Wire-backed paper lath sheets need a stable framework on which to fasten.

The metal lath is fastened to hangers from the concrete ceiling, or where lowered ceilings are required. When metal frame ceilings are to be plastered, the metal lath is tied or welded to such a base.

Metal corner beads are now used at the corners in preference to wood. Metal expansion strips are preferred to wood where expansion or divisions are required in either plasterwork or stuccowork. There are various kinds and shapes of these strips on the market as well as bases, divisional joints, and picture mold, grounds which help the application of plaster and stucco. Metal frames, channels, bracing, and lath are used extensively where plastering and stuccoing are done.

TOOLS

Clean tools are a must when doing plastering or stuccoing. Some of the tools used are covered in Chapter 2. There are many odd-shaped tools of various sizes used by plasterers. See Figs. 5-1 through 5-3.

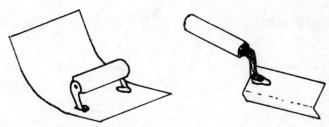

Fig. 5-1. Odd-shaped trowels.

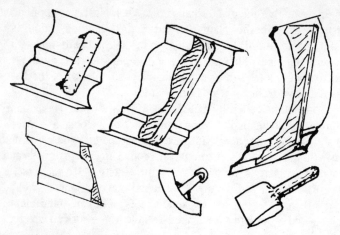

Fig. 5-2. Special molding tools.

Leveling trowels are narrower than the spreading trowels. They are longer, up to 20″ long. The best of steel blades are used in the long blade trowels.

The plaster and stucco applying and spreading trowels come in different sizes. In width they vary from 3½″ to 4″, and in length from 7″ to 20″. The blades are of stainless steel (**Fig. 5-4**).

A hawk is used to hold the mortar when troweling on the plaster mortar (**Fig. 5-5**). After the plaster mortar is put on the wall, a *darby* is needed to level off the plaster (**Fig. 5-6**).

I have used a straightedge or browning red for leveling and scraping down the green plaster. The angle plane is used on semiset plaster (**Fig. 5-7**).

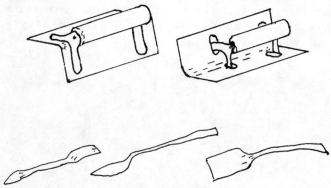

Fig. 5-3. Corner trowels and plaster carving tools.

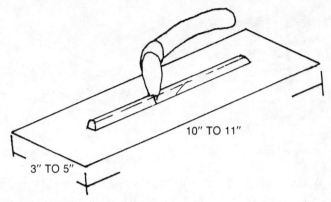

Fig. 5-4. A plaster spreading trowel.

10" TO 11"

3" TO 5"

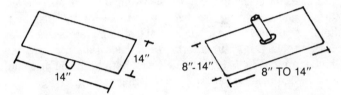

Fig. 5-5. Hawks are used to hold mortar.

14"

14"

8"-14"

8" TO 14"

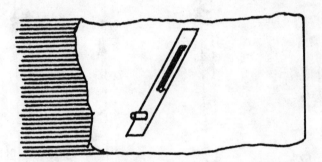

Fig. 5-6. A darby is used to level off plaster.

BLADE PLANE

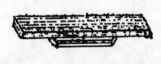

METAL LATH PLANE

Fig. 5-7. Scraping devices.

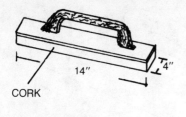

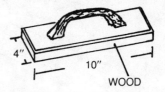

Fig. 5-8. Hand floats.

A *hand float* is a handy tool when working out the hollows or depressions in green plaster (Fig. 5-8). When corner beads are put on by the plaster mason, a long level is used.

Brushes for cleaning the tools, splashing water or splash coat, texturing, and scrubbing are necessary tools for the plasterer. Scaffolding, folding horses, ladders, and walking stilts are other necessary tools (Fig. 5-9). Mortar tables and tubs are required when plastering (Fig. 5-10). Mixing tools, both hand and machine, are used when making mortar for plastering and stuccoing.

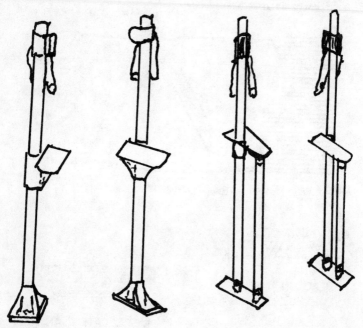

Fig. 5-9. Stilts.

Fig. 5-10. You need a mortar table and tub when plastering.

PLASTER AND STUCCO MORTARS

Plaster and stucco mortars need a binder. Some mixes require an aggregate, water, and color. These mixes make mortar that is used in various kinds of plasterwork or stuccowork.

The main binders used are lime, plaster (gypsum), cement, and magnesia. Lime is calcined limestone or chalk. Its base is carbonate of lime. Plaster is made from gypsum and has a sulfate of lime base. Cement of a gypsum base, such as some of the white cements, should not be the binder in mortar intended for use as a moisture preventative. The cement will effervesce. Any mortar mix that will not prevent moisture from affecting it should not be used outside. Magnesite is made from magnesium. When it is combined with magnesium chloride, it forms *sorel cement*, a hard and strong cement.

Some binders used in a mix make mortar which sets very hard. Others make mortar which sets soft. The mortar can be easily carved.

Binders can be used in combination. Portland cement is made from calcined limestone. Gypsum, when added to this cement, retards setting when its mortar is applied. When portland cement is used as a binder in the mix, the mortar becomes a water-preventative plaster or stucco.

Many binders can be used with or without aggregate. Manufacturer's mixing instructions should be followed. Their special uses and colorful beauty should not be overlooked.

Mortar made from these mixes can be textured. Many binders on the market require only the addition of water.

PLASTER AND STUCCO MIXES

Plaster and stucco mixes made into mortar are termed *stuff* by some masons when in a usable plastic state and in a thick, semiplastic, yet usable state. When the mortar mix is free from aggregates, it is *neat*. The mix can be in a colored or uncolored state. When the mix is conditioned with an additive, it is *gauged*. The mix is *quick setting* when it takes up and hardens in a short time, or *slow setting* when it hardens very slowly. When the mix is used as a flushing or skim coat that spreads very thinly, it can be used in combination with other materials. When it contains variable aggregates and is used as a rough coat, it is called *dash mortar, pebble dash*, or *spatter dash*. The mortar mix is called *parget* or *pugging* when stiff and used to plaster between joists to the subfloor, or roof or ceiling joints. The term plaster, as used here, applies to any mix mortar that is used for plastering, although mastics are sometimes used for plastering. The term stucco as used in this book describes all mortars that are used outside.

Unslaked or *hot* lime should be slaked well before use. It should have aged as lime putty at least three weeks before use.

One-and-a-half gallons of water can slake a bushel of average grade lime. Hydrate powdered lime should be mixed to a putty and left to stand at least 24 hours before using. Hydraulic limes, when used as a binder in mortar, give the mortar the ability to set under water. These are the best limes to use as a binder for mortar that is used outside. Lime that slakes quickly and with great heat is the best lime to use. The older lime putty is, the better, it is in a binder used for making mortar. Therefore, lime should be slaked well in advance of its use and in a quantity that does the complete job. Let it stand for at least three weeks.

Plaster of paris is a gypsum plaster. The slower setting plasters are best. When gauged with lime putty, they have a setting time according to the percentage of lime added. The percentage of gauging varies, depending on the weather and conditions under which the plaster is used, as well as the plastering time schedule.

Common cements such as *Roman, Rosendale, Windsor*, and *Medina* are best in a mortar where quick settings and expansions are desired. *Adamant* is a white cement that, when in a mortar, causes the mortar to set quickly and become hard. It is easily worked to a very fine finish. *Parian*, another quick-setting cement, sets hard to a nonporous density. This mortar should not be disturbed in its setting. A good fire-resistant cement is *Robinson*. It has other good qualities that make it a fine cement in mortars for inside plastering. Keene's cement is a white cement that, when used as a binder in mortar, allows retempering. It sets very hard, harder than any other plaster of paris base cements. Marlins cement in a mortar gives the mortar the ability to receive paint. Rosendale hydraulic cement is a first-rate outside cement binder as is portland cement. Atlas and Meduca cements are both top outside stucco cement binders. They can also be used in mortar for inside plastering.

Many cement binders may be used with lime to hasten setting and to give plastering a more hydraulic quality. Sorel cement is an exceptionally hard-setting and durable cement. When this cement is used in mortar, the temperature should not vary over 10 degrees. Expansion or contraction loosens the cement before it is fully set.

Emulsified cements and additives may be incorporated in mortars. There are also mastic cements for plastering and stuccowork.

Temperature should be watched when making and using plaster and stucco mortars. The mortars and the plaster or stuccowork should not freeze. In cold weather the material and walls should be protected with tarps or a plastic cover and heat. Heat should be even and not directly on the plastering or stuccoing work. If there is too much heat or the heat is uneven, spotty setting results (dry-outs).

The moisture content of the mortar should be adjusted to prevent dry-outs due to moisture loss or to quick, uneven setting of the mortar used. If this is not done, the work cannot be completed admirably.

Discard all lumps or other material that can cause trouble when manipulating the stuff on the wall. All setup stuff should also be discarded; if it is used, the finished job will be inferior. The mixers, boxes, and tools used should be kept clean and free of all previously mixed stuff.

MORTAR MIXES

Lime plaster mortar for the scratch or first coat should be mixed well. The hair or fiber should be separated and evenly dispersed throughout the mix. The water content should be enough to make the desired consistency for the mortar.

First Coat of Lime Plaster Mortar

1 cubic foot lime putty.
2 cubic feet very fine sand.
One-half pound chopped hair or fiber.
Water.

Second Coat of Lime Plaster Mortar

1 cubic foot lime putty.
3 cubic feet very fine sand.

1/5 cubic foot fiber.

Water as required.

Finish or Third Coat of Lime Plaster Mortar

1 cubic foot lime.

1/5 cubic foot gauging plaster.

1½ cubic feet while cement or 1½ cubic feet Keene's cement.

Coat Mixes for a Sand Finish

Should a sand finish be desired, the finished coat mix should be one of the following mixes:

1 cubic foot lime putty.

1/10 cubic foot gauging plaster.

½ to 1¼ cubic feet very fine sand.

Water.

Color if desired.

1 cubic foot lime putty.

1 cubic foot Keene's or white cement.

4 cubic feet very fine sand.

Water.

Color if desired.

When dry powdered limes are used, the amount of lime should be from 5 to 10 percent. Should pumice, *perlite*, or waterproof sawdust be used instead of sand, the volume measure is the same.

Gypsum plaster mortar mixes are as follows.

First Coat of Gypsum Plaster Mortar Mix

1 part plaster (fibered).

1 part fine sand.

Water.

or

1 part unfibered plaster.

2½ to 3 parts very fine sand or light aggregate.

Water.

For the second coat, the preceding formula mix can be used.

Finish Coat of Gypsum Plaster Mortar Mix

1 cubic foot unfibered plaster.

2 cubic feet very fine sand.

Water.

Color if desired.

Prepared Plasters

Prepared plasters may be used without the addition of sand or aggregate. Where fire resistance is desired, less sand should be used in gypsum cement plasters. It is recommended that a sealer coat be applied. This sealer coat should be set well before gypsum plaster mortar is used. Portland cement grout makes a good sealer and bonding coat when dashed and worked into the base with a stiff brush. It should be kept damp for two days, and then allowed to dry before the plaster mortar is applied.

There are a number of prepared plasters such as Universal, U.S. Gypsum, Red Top, and many float finishes. If directions are followed, they produce material for top notch plastering.

Cement stucco mortars covered in this book are the best outside mortars. They are water-repellent and can be used on inside plastering and especially on places where moisture is present.

First Coat of Cement Stucco Mortar

1 part portland cement.
3 parts fine graded sand.
Water.
or
1 part portland cement.
⅓ part hydrated lime or lime putty.
3 parts fine graded sand.
Water.

Second Coat for Cement Stucco Mortar

1 part portland cement.
¼ part hydrated lime or lime putty.
3 parts fine graded sand.

Finish Coats for Cement Stucco Mortar

1 part portland cement or white cement.
2¾ parts very fine graded sand.
If desired, add some coarse aggregate.
Water and not over 3/100 part waterproofing or damp-resistant powder or liquid, if desired.

Color may be added to a finish mix to not over 10 percent of the binder.

White cement is best when color is added.

or

1 part white cement (Atlas, Meduca, or other).

1 part Ohio hydrated lime.
3 parts very fine white sand.
Water.
Color.
5 percent part other larger aggregate if desired for some finishes.

Rough Cast Finish

A good rough cast finish that can be the finish coat is as follows:
1 part portland cement.
2½ parts fine sand.
¾ to 1½ parts aggregate depending on size.
1/50 part waterproofing material if desired.
1/75 part color (should always be a mineral color).
Water.

Splash Coat

There are splash coats that are splashed on the finish coat. A good splash coat is as follows.
1 part cement, portland or white.
1/10 to 1/4 part hydrated lime.
¼ to 1 part very fine sand.
Color.
Water.
A small percentage of other aggregate that can pass through a ⅜" screen may be added for pebble finish.

Molding Mortar Mixes

The following molding mortar mixes are excellent.
1 cubic foot gypsum plaster.
22 pounds molding plaster.
2 cubic feet very fine sand.
Color.
Water.
or
1 cubic foot gypsum plaster.
½ cubic foot Keene's or other white cement.
1½ to 2 cubic feet very fine sand.
Color.
Water.
A good outside mix is as follows:
1 cubic foot portland cement.

1/5 cubic foot hydrated lime.
2¼ cubic feet very fine sand.
Color.
Water.

The addition of 10 percent part plaster of paris to the binder of this mix gives the mortar a quicker set for molding work with templates, molding tools, etc. These tools should be coated with *linseed oil* before use.

APPLICATION AND MANIPULATION OF PLASTER AND STUCCO

Many plaster and stucco jobs are two-coat work. The finish or second coat varies in material mixes as well as in appearance for different jobs. There are one-coat jobs performed by plastering or stucco machines and artificial coats with various paint concoctions. A truly first-class plasteringwork or stuccowork is a three-coat cover. The first or scratch coat is laid as a cleating or rendering the base that holds the second and third coats to the wall. The second coat or brown coat is a filler and strengthening coat. It is on this coat that trueness and leveling is done. If the second coat is finished and no other coat is applied, the job is known as a two-coat job of plasterwork or stuccowork. The third or finish coat is laid on for appearance and protection. Three-coat work is sometimes designated as lath, lay, float, and set, or, when applied to a base requiring no lath, as render, float, and set.

The plasterer or applicator must see that there is a proper base on which to put plaster or stucco. The base on which cover mortar is laid in many cases is very uneven, especially some masonry bases. Plaster and stucco grounds must be used. These grounds can be of wood or metal. The plaster mason can make the grounds with mortar should it be desired, especially away from corners and openings. When these mortar grounds are made, a straightedge is used to true them.

Metal grounds also require a straightedge or a line. The wood grounds can be removed and their place filled in. The grounds at the base can be left on as a ground nailing-back for the baseboards; the wainscoting grounds can also be left on.

All grounds should be placed true to the finish coat. The other coats can be recessed for their required depth, leaving room for the finish coat. The brown coat should be made from ⅛″ to 1/16″ below most finish coats. The finish coat does not require grounds.

All mortar mixed should be fitted for the base or wall to which it is applied. Masonry bases should be well chipped or picked if smooth. Joints should be raked out, and all loose material should be removed. All bases should be cleaned and be free from grease, oil, and paint. If necessary, the base should be washed down with 10 percent solution of acid (1 part muriatic acid to 9 parts water). The items to use for this wash are a pair of rubber gloves, a stiff brush, and a rubber, plastic, or enamel pail. The acid wash should be washed off with water before mortar is laid.

The masonry should be sprayed with water to help eliminate suction and stop dry-outs of the mortar cover. A wood lath base also needs spraying. Metal lath and fabricated bases do not need spraying. Some mortars require a galvanized metal lath when this type of lath is the base. Screeds such as corner beads should be used and are a help at all corners and openings.

First Coat

The first coat is laid on evenly with the trowel. If the spraying, especially on masonry, does not stop suction, a splash coat should be applied of neat cement (cement and water to consistency of half-and-half cream).

The stuff is laid on to whatever depth necessary to have a somewhat even surface. The work entails the taking of the mortar from the mortar table or tub with the trowel to the hawk (Fig. 5-11).

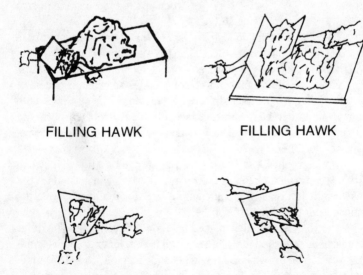

FILLING HAWK FILLING HAWK

FILLING TROWEL FILLING TROWEL

Fig. 5-11. Note how the mortar is handled.

It is then applied to the base. The method of handling the trowel is easily accomplished with practice. Some masons snatch the mortar from the top down with the trowel. Others catch the mortar from the bottom or scrape it along the tabletop with the trowel to the hawk. How the mortar gets on the hawk does not matter; how it gets from the hawk to the base does matter (**Figs. 5-12 and 5-13**). This

Fig. 5-12. Plastering a ceiling.

procedure requires a bit of practice as mortar has the tendency to slip from the trowel. The trowel spreads the mortar in an upward and outward motion on the base. It is best to start at the top of a base at the edge of a ceiling (from one side to the other on the ceiling). The mortar is pressured to the base just enough to have the described thickness of mortar on the base. Enough pressure should be used to insure mortar cleats which hang on metal lath, wood lath, and joint rakeouts. A mortar clinching key helps mortar stay put. When the stuff is laid on masonry, the thickness varies (**Fig. 5-14**).

In poor-class plastering or stuccoing, this first coat is floated down. No other coat is laid. In a first-class job, this coat is darbied to trueness. Just before it takes a firm set, it is scratched. Scratching loosens the stuff if it is not firmly set to the base. Do not scratch it too soon. The stuff should be scratched from 1½" to 3" apart. Deep scratching in many cases subjects the plasterwork or stuccowork to cracking.

The first coat should be almost dry before the second coat is laid. On some bases this first coat may need spraying with water, by the use of a brush or fine spray nozzle, to eliminate suction. Too much suction draws the moisture out of the second coat before it can be properly laid and worked. Too fast drying causes the stuff to dry out to a powder and be useless. I find that a saturation with alum water helps restore chalklike spots.

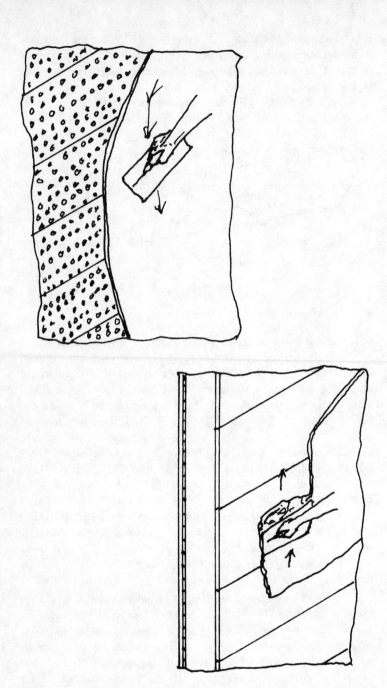

Fig. 5-13. Use the trowel properly.

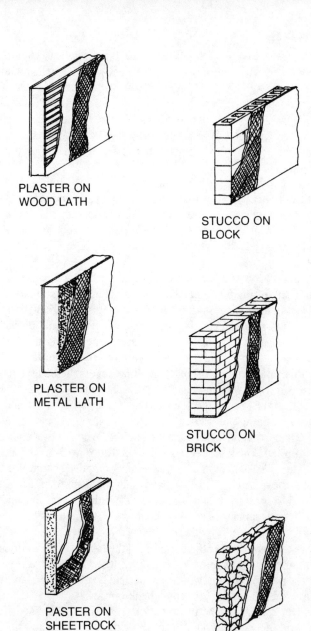

PLASTER ON
WOOD LATH

STUCCO ON
BLOCK

PLASTER ON
METAL LATH

STUCCO ON
BRICK

PASTER ON
SHEETROCK

STUCCO ON
STONE

Fig. 5-14. Plaster bases.

Second Coat

Before applying the second coat, the first coat should be scraped down to remove any loosely hanging stuff. Then it should be brushed. It should be wetted to eliminate suction if necessary. The stuff is now troweled on to the first coat to a depth of ⅜". It should be packed well and screeded to within 1/16" to ⅛" of the grounds of a three-coat job, or to the screeds if a two-coat job. It should be darbied, using water, to a perfect flatness and free from catfaces (depressions).

The darby is used to bring the second coat to a true flat surface in both horizontal and perpendicular trueness by circling, criss-crossing, and figure eighting with the darby, float, or rod. Water is splashed on which helps to fill in the catfaces. If spot nails or spot grounds are used, they should be removed. The holes should be filled.

A good way to insure true flatness, with less chance of the second coat setting up before flatness is maintained is to use spot grounds. Should mortar grounds be used, they should be within the reach of the screeding straightedges or rods. They could be run at the inside corners and edges of the base. Corner beads are best at openings and corners. Spot nails driven into the required depths, spot lath, or mortar spot grounds are time-savers. They give more time to work the second coat to a true flat finish. The spot nails should be driven in after the second coat is laid. The spot lath should be removed. Scrape off the spot mortar. All holes should be filled with fresh mortar and leveled true to surface. When the second coat is taken up and firm, it is scoured and floated down with hand floats.

Water is splashed on when needed. A circular or figure eight motion is best when a float or sand finish is desired. First-class work is floated down at least three times within three-hour intervals. The last scouring is done without water when the second coat is practically dry. This gives a roughness that is a better base for the third coat. Temperature affects timing of the takeup of the stuff, and certain procedures should be performed when the stuff is ready. The three floatings not only pack, shape, and level the stuff, but they help insure the finished job from check cracking.

In two-coat work, the plasterwork is considered finished after the first or second floating. No finish or third coat is laid on.

When the mortar stuff is a cement mix, hand floating is sometimes dispensed with, and a finish coat is laid before the brown or second coat is dry. It can be laid on as soon as the brown coat is firm. Some cement mortar stuff sets nearly as fast as it is applied or laid.

Third Coat

The third coat is a hard set coat or a putty slick coat. This coat is also a finish coat and can have many different textures formed with different mixes and workmanship. The stuff is laid on the second coat with a trowel to a depth essential to the finish desired. This is the coat that protects, beautifies, and dignifies the art of plastering. This coat is applied and manipulated various ways according to the finish desired.

The hand troweled finish is most generally used when the plasterwork is to be painted with various color concoctions. The stuff is laid on and floated to a true flat surface. Where suction interferes, the stuff is double-laid, which seals off suction. Because this coat is a thin coat, a good practice is to spread a trowel of the stuff, from bottom to top, at the upper part of the wall in one sweep of the trowel. Lay the joint down with a downward sweep of the trowel. This method can be reversed sometimes to advantage. Should the trowelful run out on an upward sweep, the shortage has to be dubbed in, necessitating extra work. The stuff is generally laid on with an outward motion. The joint is laid down with a comeback downward motion of the trowel. This stuff is allowed to take up when true, flat, and waveless. It is finally scoured with a hand float and water and then scoured again with water to work out the catfaces. The third time it is floated with a square edge trowel which cuts off all edges. Water may be used sparingly as it is brushed down.

After an even close-grained surface is obtained, troweling should be done. A polishing trowel (an even-edge trowel) is best for this purpose. Water is used when troweling. Troweling should be done crossways and up and down until the color is even, if colored stuff is used. The surface should be free from fatty glut. Near the end of troweling, the water should be sparingly used. The last part of troweling should be up and down. The hard finish coat should be lightly brushed with a long-hair brush. This method insures first-class three-coat work. An inferior finish coat has just a thin skim coat of quick-setting stuff laid on and is floated as lightly as possible. It is brushed down as soon as it is laid. Gauged stuff should never be scoured or troweled when set, as it peels off.

FINISHES

There are many plastering and stuccoing finishes that can be applied to the first, second, or third coats. Certain finishes fit well with certain architectural periods. Some finishes are manipulated with the finish stuff. Others are added to the finish coat, and still others are carved out of the finish coat.

American Finish

The *American finish* is textured with a float used up and down on the finish. A brussels carpet-covered float leaves a sanded finish. A float placed on the finish stuff after darbying, and quickly pulled off, leaves a suction-textured finish. A sponge used in the same manner leaves a *stippled* finish. A whisk broom or stiff brush used vertically and horizontally on the finish leaves a basket-textured finish. This finish work is done before the finish whether colored or uncolored, sets (Fig. 5-15).

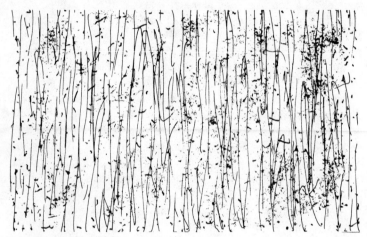

Fig. 5-15. American finish.

Spanish Finish

The *Spanish finish* is accomplished with a brush or the palm of the hand before the finish stuff is hard set (Fig. 5-16).

English Finish

The *English finish* is done with a snub-nose mixing trowel by dabbing different colored stuff on the finish coat after it has taken up and before it has set. Suction should be eliminated before this colored stuff is slapped on (Fig. 5-17).

Pebble Dash Finish

The *pebble dash finish* is thrown on the finish where it is in a semiplastic state. This dash may contain various sizes and kinds of

Fig. 5-16. Spanish finish.

Fig. 5-17. English finish.

aggregate averaging ¼″ to ½″. A coarse-bristled brush or heavy whisk broom is forcefully dipped into the pebble dash grout mix. The stuff is thrown on the semiset finish coat with force. This is repeated until the stuff is evenly splashed (Fig. 5-18).

Dry Rough Cast Finish

This *dry rough cast* finish is applied the same way as the pebble dash finish. The aggregate is dry. A board is used to brush it from an open end box with force. The finish stuff should be freshly laid and darbied. The aggregate may vary in size and color. This material and the pebble dash stuff can be thrown on by hand, or a machine can be used (Fig. 5-18).

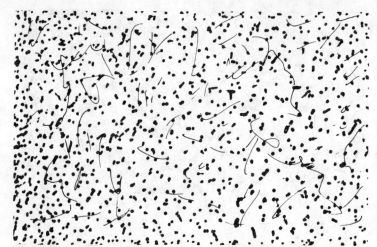
Fig. 5-18. Pebble dash finish.

Travertine Finish

The *travertine finish* is accomplished when the finish is semiset. The stuff is splashed on horizontally with a long-handled brush. This leaves a streak of holes in horizontal lines as small pockets or suction cups. A trowel is used to lay down the stuff, but not to the extent of filling the pockets. Another method is to stipple horizontally paths or furrows in the finish with a stiff brush before lightly troweling. There are also other methods (Fig. 5-19).

Fig. 5-19. Travertine finish.

98

Smooth Trowel Finish

This finish requires only trowel finish after darby work (Fig. 5-20). All these finishes, except the pebble dash and rough cast, may

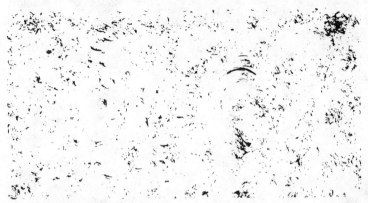

Fig. 5-20. Smooth trowel finish.

be joint-grooved like a stone bond. The grooves can be filled with a different colored material to give the appearance of mortar joints. Joint cutting or grooving is accomplished with hacksaw blades fastened together on a board. Work it back and forth to make the groove. After the joint stuff is set, it can be sanded by hand or by power, should a smooth finish be desired. Otherwise, the semiset joints should be tooled with one of the many jointers shown in Chapter 2.

Textured finish coats can be colored if desired. Various colored mixes may be carved, and various coats can be cut and peeled

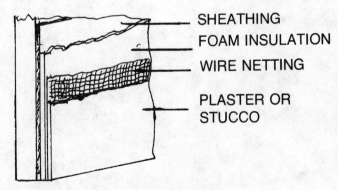

SHEATHING

FOAM INSULATION

WIRE NETTING

PLASTER OR STUCCO

Fig. 5-21. Stucco can be applied on foam insulation.

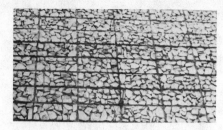

Fig. 5-22. Stucco cut like stone.

Fig. 5-23. Lined stucco.

Fig. 5-24. Application of stucco.

off to the desired shape and color. This creates a variety of many-colored designs. Lime or gauged finishes may be stenciled or painted with water colors (not vegetable colors) when they are semidry. These designs are permanent.

These finishes can be ruled, marked, or cut to have the appearance of tile, block, brick, or stone. An artisan can actually duplicate the shape and color of such construction units. These kind of finishes are decorative stuccowork as well as protective coats.

These are masonry covers such as brick veneers, stone veneers, black veneers, tile veneers, plasters, and stucco covers. These are protective coats that protect the surface of the structural parts. They are beautiful, also.

Stucco is a protective coat. It can be applied on foam insulation (Fig. 5-21).

Figure 5-22 shows stucco cut like stone. Figure 5-23 depicts lined stucco. Figure 5-24 shows the application of stucco.

Chapter 6

Cement Finishing

Cement finishing comprises not only the top and wall surfaces to be finished, but also the supervision of pouring, finishing, and cleaning cementwork. In order to secure a good cement mix, the ratio of the binder to the aggregate used must be considered. Methods of making, handling, and finishing are taken into consideration when a top job of cementation is required.

Early strength cement, when used, is in the same proportions as regular portland cement. Concrete made with early strength cement develops nearly the same strength 1½ days from its mixing as does regular portland cement concrete in 8 to 10 days; thus, forms may be stripped earlier.

Air-entrained cement is also extensively used. Concrete produced with it has a percentage of minute air bubbles throughout its cured bulk and is far more workable. This cement can be produced with different percentages of air-entrained content. The air-entraining agent, though, can be obtained separately and added to regular cement. Air-entrained concrete is more resistant to salt action and frost than normal concrete. When using this cement, the water quantity for mixing can be reduced considerably where the air percentage is 4, 6, etc. Sand quantity in the mix can be reduced 4 to 6 pounds per 1 percent of air. The sack of 1 cubic-foot unit measure is the same as with other cements. The air-producing agent increases this cement's volume. Generally, in rich mixes the water requirement should not be reduced.

Airocrete (containing a chemical gas-producing agent causing it to expand, even to the point of reducing the weight of the concrete is to as low as 43 pounds per cubic foot) is another cement used extensively. Its compressive strength may be as low as 500 pounds per square inch in light-weight concrete.

Lightweight concrete has always been made by using lightweight porous aggregates. More water is used with these aggregates. They have to be sprayed well with water before they are incorporated into the mix.

All mortar and concrete may be colored by adding color to the dry binder. Colors must be well mixed and not exceed 12 percent of the binder quantity. Mineral colors are best, but there are also good fade and weather-resistant colors on the market. The mix should not be sloppy or watered too heavily when a color is used. Other ways of coloring concrete are described later.

Additives can be used in the mix for durability. Each additive has manufacturer's directions for mixing, handling, and use.

On small jobs such as patchwork, the mixing can be done with a bucket trowel. Tucking work generally takes small batches that are made by hand with the hoe or shovel. When several masons are tucking, a small machine mixer is used. Mixing fireclay may require a paddle and bucket (no aggregate, only fireclay and water mixed to a creamlike consistency). On large jobs a power-driven mixer is used.

CONCRETE MIXES

A good mix for footings and foundations is a proportion of 1 part cement, 2 parts sand, and 4 parts aggregate. A 1 part cement, 3 parts sand, and 5 parts aggregate proportion is as good as a 1 part cement, 3 parts sand, and 6 parts aggregate mix for making concrete. The latter two mixes do not make watertight concrete, but they may be used where water is not a problem.

A 1 part cement, 2½ parts sand, and 3 parts aggregate proportioned mix makes concrete that can be used for footings and heavy walls such as bearing walls, piers, pavements, lintels, etc. A 1 part cement, 1½ parts sand, and 3 part aggregate proportioned mix makes good concrete for footings and heavy walls that are protected from alternate drying and wetting.

A 1 part cement, 1½ parts sand, and 2½ parts aggregate proportioned mix, having 10 percent very fine sand, makes good, watertight concrete. A 1 part cement, 1½ parts sand, and 2 parts aggregate proportioned mix is the best for making material for reinforced concrete where high loads and heavy use is expected.

The 1 part cement, 2½ parts sand, and 3½ parts aggregate proportioned mix is used to make a good concrete for most concrete structures. This concrete is safe in all moderate exposures to the elements.

The amount of water used in the mixes governs the strength and quality of concrete. Compressive strength of concrete such as 2,000, 2,600, or 4,000 pounds per square inch at 28 days varies greatly according to the amount of water in the mix. Less water enhances the strength of all concrete.

A 1 part cement, 2⅓ parts sand, and 3⅓ parts aggregate proportion of well-graded aggregate, with not more than 6¼ gallons of water, makes a good, durable concrete that withstands weathering. Compressive strength of this concrete can be regulated by the amount of water used (including surface water in the aggregate). When 5 gallons of water are used, the concrete has a strength of 3,800 pounds per square inch. When 6 gallons of water are used, the concrete has an approximate strength of 3,000 pounds. When 7½ gallons of water are used, the concrete has an approximate strength of 2,000 pounds in a 28-day test.

A 1 part cement, 3 parts sand, and 6 parts aggregate proportioned mix (cement, sand, slag, or cinders); a 1 part cement, 4 parts sand, 7 parts aggregate proportioned mix; or even a 1 part sand, 4 parts sand, and 8 part and aggregate proprotioned mix can make a good concrete for a floor fill. Another good lightweight floor fill, one that can be nailed to, is a 1 part cement, 2½ parts sand, and 6¾ parts aggregate proportioned mix.

A lightweight concrete for floor and roof fills is 1 part cement, 2 parts sand, and 8 parts aggregate proportioned mix (cement, zonolite, or heat and steam-expanded vermiculite) with not over 4 gallons of water used. Compressive strength varies, depending on the amount of aggregate used, from 140 to 580 pounds per square inch. Again, the amount of water affects its strength.

A 1 part cement, 3 parts sand, and 1½ part aggregate proportioned mix (cement, fine-grade pumice, and coarse pumice), using 4¾ gallons of water for aggregate saturation and up to 4 gallons of water for mixing, makes a good lightweight concrete with an approximate strength of 2,500 pounds per square inch. Should the aggregate be doubled, the strength is approximately 1,000 pounds. The weight is reduced from approximately 75 pounds to 55 pounds per cubic foot.

Other types of concretes are made with fibrous aggregates such as sawdust, pulp, etc. Stones dropped in a concrete mix should not be closer than three-fourths their thickness.

MIXING CONCRETE

If the cement mix is made by hand, a screen, hoe, shovel, mortar box, a 1-cubic-foot box or measure, water barrel, measuring bucket, and two or three more buckets are needed. If color is used, another box may be needed.

First, the aggregates are sifted and graded (a long-handled shovel is best for this job). Sand is shoveled up and thrown, scattered, against the slanted upright screen. As the sand slides down the screen, usable sand passes through. The remaining unusable sand is thrown aside as it accumulates. Coarse aggregate is also graded this way if it is not bought properly graded.

The next step is to place the right amount of sand into the mortar box and level it off with the hoe. The required amount of cement is placed on the sand and leveled off with the hoe. Use the hoe to chop dry cement and sand together into a uniform, well-mixed mass. If color is to be used, it should always be mixed well with the dry cement in a separate box before adding the cement to the sand. The water can be added slowly. Mix the mass into a plastic, pliable mass. Then add the coarse aggregate, which is well mixed into the plastic mass. All of it must be well mixed together while using the required amount of water.

When using a power mixer, the procedure is different. Pour in some of the mixing water, add a shovelful or two of course aggregate, a couple of shovelfuls of sand, and then the required amount of cement, water, sand, and coarse aggregate. This method of loading the concrete mixer generally keeps it from clogging or dry packing the mix. Well-mixed concrete requires at least two minutes of mixing after all materials are in the mixer. A 3-minute mixing time is best.

When making a lightweight concrete, always saturate the porous aggregate with water before the cement is added. This procedure stops the aggregate from absorbing water from the cement binder, which is needed for chemical action. Saturation also stops the aggregate from floating.

When mixed, concrete is placed. Any work necessary should be accomplished quickly before the concrete begins its initial set. Disturbed setting produces inferior concrete.

The weather must be considered when making concrete mixes. Ice should never be in the mix. Concrete should be protected from frost in freezing weather. The aggregate and the water may require heating to make safe concrete. Water should not be boiling when used. Concrete temperature should not be less than 40 degrees Fahrenheit nor more than 80 degrees Fahrenheit. The concrete should be protected from extreme weather conditions at all times until it has reached the safety zone in its curing stage. In rainy weather it should be covered, as excess moisture before its initial set may impair strength and quality. Hot weather dries out fresh concrete too fast, making it porous and dry, and thereby reducing its strength.

All tools and equipment should be kept clean and in top working condition. This also helps when making uniform quality mixes. Some tools will not work properly if caked or clogged with dried or set mixes.

It is generally more economical to buy ready, transit-mixed concrete delivered fresh to the site. Any mix will be consistent, quality-controlled, and reasonably priced. The cement mason supervises and handles this mix when it is delivered. This holds true whether the mix is delivered from a central bulk plant by pipe, trough, chute, wheelbarrow, hand or powered two-wheel cart, track, or crane bucket.

PLACING AND FINISHING CONCRETE

During a pour (placement), a comealong (a long-handled, variable length, curved plate, hoelike tool) is used to push or pull the mix to fill in and even it to near grade. Wade out in the mix when grade grounds are far apart. If steel stakes or pegs driven to grade are used instead of grounds in the intermediate area, help may be needed to level the pour with long screeds.

After screeding the mix to grade, it is tamped or the *roller packer* is used to settle the mix and seat the larger aggregate below the surface (Fig. 6-1). They will not be needed if grounds are placed to accommodate the vibrating power screed or rod. Such a screed may be 30' long, a motor producing 3,000 to 7,000 vibrations per minute (vpm). Vibrating screeds may be larger or smaller. A concrete cement mix should never be overvibrated.

Concrete as a base for other materials requires a finish that best accommodates the topping. If a rough finish is needed, the pour should be checked for evenness of grade. If it appears fatty or smooth-textured, it should be roughed with a rod, screed, or rough

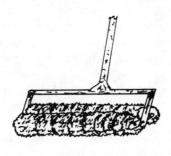

Fig. 6-1. A tamper and packer.

float after the water has disappeared. This insures a better amalgamation with the topping material.

Concrete for a sidewalk, steps, porch floor, pavement, or driveway requires a fine-textured, water-impermeable roughness. It should not be smooth enough to cause slipping. After the mix has been screeded or rodded to grade, and the surface water has practically disappeared, the concrete mix should be floated down to an even texture.

Floats are of various sizes and materials; some require two workmen to operate. Long, wide floats reach a foot or so beyond grade grounds (pavement forms). They are worked back and forth the length of the pavement by a man at each end. A float can be 2″ × 12″ plank, a wide fabricated belt, or even a garden hose. Take care not to let the float gouge out the mix. If a metal float is used, the concrete cement pour should be broomed to give it the desired rough finish.

For concrete cement requiring a smooth finish (e.g., for a basement floor), after floating or darbying, a metal float and a trowel are used. A power-driven, metal-bladed cement finishing trowel can be used instead of the hand trowel to finish the concrete element. All work on the mix, except the troweling, should be done before the initial set in order not to weaken the final cured concrete element. Finish troweling can be done before its hard set. All concrete, for whatever use, requires attention through the curing stage to secure the best, most durable concrete.

Colored Concrete

Colored concrete can be made by two methods—either

107

method is cheaper than coloring the mix when it is made. In one method, use this dry color mix: 1 part cement, 1 part sand, 1½ parts aggregate, and not over 10 percent color (1 part cement, 1 or 1½ parts very fine clean sand, and 9.4 pounds color). The color is thoroughly mixed with the cement before sand is added. When the color mix is ready, it is applied (dusted) uniformly over the poured mix after it has been screeded and brought to a true, even grade. The color mix is then floated until it becomes saturated with moisture. When the surface water has disappeared, floating or troweling is resumed to obtain the desired finish.

In the other method, a wet cement mix with color is prepared: 1 part cement; 1, 1½, or 2 parts sand; and not over 10 percent color (1 part cement; 1 part very fine sand; 1½ or 2 parts coarse graded aggregate—⅜ to ¼ size, No. 8 or No. 4 screen—and 9.4 pounds color). The maximum size of the coarse aggregate is governed by the desired depth of the top color coat. This wet mix has not over 5 gallons of water per cubic foot or sack of cement used.

A subbase pour is broomed off with a stiff-bristle broom, wire broom, or brush to free the base slab of loose materials, aggregates, and *laitance*. The colored mix is placed, screeded to the correct grade, and darbied or floated. Grooving and edging is performed with hand tools (Fig. 6-2).

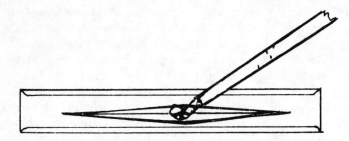

Fig. 6-2. The darby is a useful tool.

Unless a power groove is used, the handmade groove should be checked and corrected or rerun before the slab is hard. This is the last work performed with the cement tools. If the pour is too large for one mason to handle, more masons may be needed to properly finish the work.

By grooving with a power saw, the shape of stone, brick, tile, or block can be duplicated. Colored cement can be put in the grooves as a mortar joint.

Forms and Slabs

In a foundation or any form fill the placing of the mix should be watched. The pouring should be stopped if there is danger of the forms expanding or giving way. A mix exerts pressure on forms in proportion to the weight of the mix per cubic foot and depth of pour, as well as to the temperature of the mix and length of time to fill the forms. If the mix is poured fast to a depth of 10', and is still in a semiliquid state, pressure at the bottom of the forms would be equal to the weight of the mix times the height (in feet) of the form. If the weight per cubic foot is 150 pounds and the temperature of the mix is 64 degrees Fahrenheit, this weight pressure could be 1,500 pounds. If the temperature of the mix is higher, the weight during the course of average pouring is less, and the mix takes up more readily. If the temperature of the mix is lower, the weight pressure is more. If vibrated, the mix holds its pressure during vibrating.

Forms should be filled properly. The concrete pour should not be overvibrated. Pockets or honeycombs should not be in the poured mix. The inside of the form should be free of splashes or spills which, if dry before pouring, can disfigure the finished concrete wall. A mix does not adhere to magnesium forms or adhere readily to wood forms treated with form oil. Some ready-made forms also need oiling. Reused forms should be clean.

At the end of the day's pour, the concrete pour should be rechecked and repaired if needed. The templates and bolts should be checked. The forms, ties, bracings, etc., that were installed may need attention (Fig. 6-3). (Form building is not taken up in depth in this book.)

The finished pour may require floating down. Pours unfinished at the day's end should be left rough, stepped down, and free of laitance if possible, leaving a better unfinished pour to knit to when pouring is resumed.

If the pour was in depth (e.g., a foundation), steps should be formed in the pour. The poured mix should be free of loose material and laitance. Grout the unfinished pour just before the next pour is made. This grout can be thin and neat (rich in binder), without the addition of sand in the grout mix, and should be brushed in.

A slab can be poured in layers, in different thicknesses, colored or not, and topped neat (richer mix, colored or not). The slab can be poured in stages, generally leaving off the pouring at an expansion joint or where a groove joint is; sometimes a pour is stopped at an intermediate point. A one-layer slab is shown in Fig. 6-4.

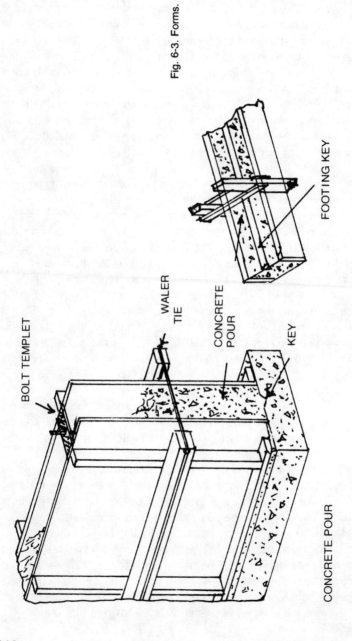

Fig. 6-3. Forms.

BOLT TEMPLET

WALER

TIE

CONCRETE
POUR

KEY

CONCRETE POUR

FOOTING KEY

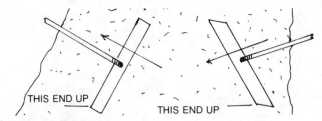

Fig. 6-4. A one-layer slab.

Masonry Floors

Footings on which a foundation stands can be reinforced with steel. Foundations should, if possible, be poured continuously to completion. Subbase bearing floors should be of a mix with hard natural aggregates. Should this structural base pull a shrinkage, strain, or vibratory crack, the top finishing material, if not structurally separate, is likely to crack in the same place. The fill, or second pouring resting on the substructural base, may be a lightweight aggregate concrete, possibly reinforced with welded reinforcing wire. The top finish slab can be of aggregate concrete—porous, lightweight, or fibrous aggregate concrete. It can be of terrazzo, mastic, asphalt, tile, or wood. In one form of construction, the structural section incorporates structural tile and reinforcing, including fill and slick-finished floor (Fig. 6-5).

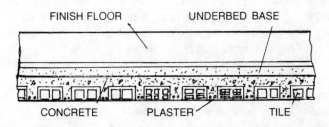

Fig. 6-5. Diagram of a masonry floor.

Another form of construction incorporates inverted U-shaped metal pans, reinforcing, and steel-ribbed metal lath ceiling, which is hung for plastering (Fig. 6-6). The finished floor or roofing, as required, may be installed on the lightweight fill atop the hard aggregate concrete-bearing structure.

A structural subbase floor and beam, one-piece pouring, may be topped by a lightweight aggregate fill. This can accommodate any kind of finished floor.

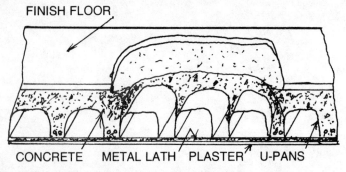

FINISH FLOOR

CONCRETE METAL LATH PLASTER U-PANS

Fig. 6-6. Inverted U-shaped metal pans are used for this construction.

A floor base of this kind should be reinforced along with the beam construction of which it is a part. The base should be watered down well and grouted if the finished floor is to be a part of it. This is not necessary when a sand bed is used (Fig. 6-7).

A mosaic terrazzo finished floor may rest on a lightweight concrete reinforced with welded wire. This lightweight fill tops a structural hard aggregate concrete slab and beam that has light-weight soffit blocks in its reinforced construction.

A sand cushion bed can also be used topped with an underbed on which to lay the terrazzo finish floor. This method saves the finished floor from having any structural cracks that may appear from the concrete bed below. If sand is not used on which to put the terrazzo finish bed, a good grouting should be done on the structural bed. The structural beams and bed should be examined before a finish floor is started (Fig. 6-8).

Figure 6-9 depicts another kind of subfloor. The reinforced concrete is poured with tile as a spacer for lightness. Plaster is run right on the bottom of the concrete and pumice. Lightweight block sand is placed as a separation to evade the cracking of the finish floor. The terrazzo brass strips are placed on the underbed of the terrazzo finish. This method of construction requires a solid deck on which to pour the subfloor.

Roof Construction

Roof construction can be handled in the same way as floor construction. The concrete floor should be reinforced as well as the concrete beams, whether they are poured with the floor or separately. A lightweight underbed may be used. The finish floor should be topped with a built-up roof as required (Fig. 6-10).

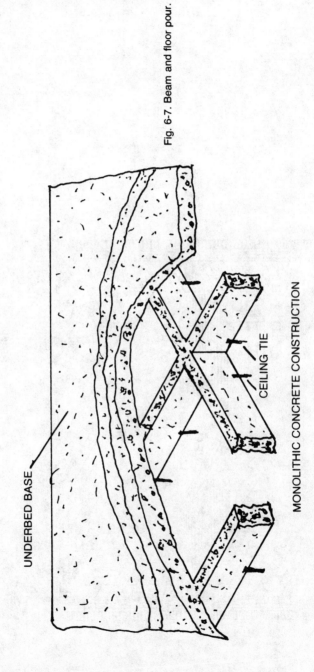

UNDERBED BASE

CEILING TIE

MONOLITHIC CONCRETE CONSTRUCTION

Fig. 6-7. Beam and floor pour.

113

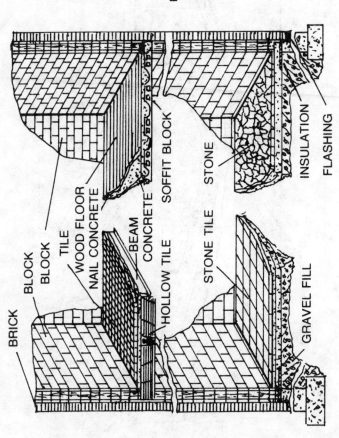

Fig. 6-8. Masonry combinations.

BRICK
BLOCK
BLOCK
TILE
WOOD FLOOR
NAIL CONCRETE
BEAM
CONCRETE
SOFFIT BLOCK
HOLLOW TILE
STONE TILE   STONE
GRAVEL FILL
INSULATION
FLASHING

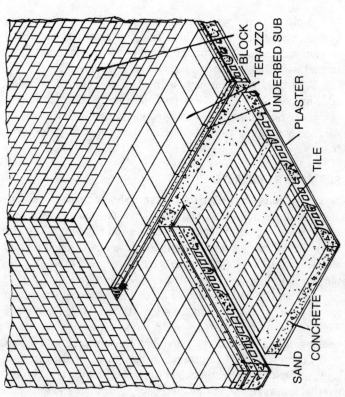

BLOCK

TERAZZO

UNDERBED SUB

PLASTER

TILE

SAND

CONCRETE

Fig. 6-9. Masonry construction layout.

115

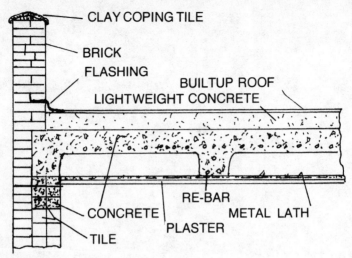

Fig. 6-10. Masonry roof construction.

A pitched roof, steel frame truss construction may be topped by lightweight nailing concrete. This concrete is reinforced with welded wire, cable-incorporated, and is poured on metal lath and topped with shingles nailed directly to it (Fig. 6-11).

A pitched roof can have a tile roof in place of shingles. A built-up roof is not the best roof, unless the pitch is less than one-sixth. Even then, the asphalt topping runs down in hot weather. All roofs and coping should be flashed for water seepage and protection.

PROTECTING CONCRETE

Any cement mix used for concrete should reach its initial set, continue its setting without interruption while being worked on, and come to a set state of desired hardness and durability. The curing stage actually never ceases until concrete has reached its climax and deterioration sets in. Properly cured concrete has a long useful life.

During the early period of its curing, concrete must be protected from frost, wind, and sun—any of which will stop its curing. This danger period lasts 5 to 18 days, depending on weather, composition of the concrete, and how the concrete was mixed.

Concrete should be protected throughout its danger zone to prevent dry-outs or heat-cracked concrete. Frostbitten, chilled, or overheated mixes should not be used. Pouring cement concrete

116

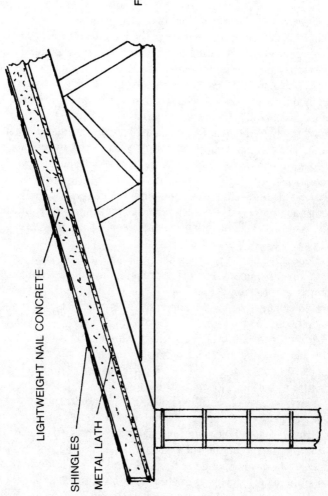

LIGHTWEIGHT NAIL CONCRETE

SHINGLES

METAL LATH

Fig. 6-11. Lightweight concrete.

117

should not be done if the temperature is above 100 degrees Fahrenheit or below 40 degrees Fahrenheit, unless the temperature is under control by shade or heat. A pour should be covered and kept damp for at least 7 to 12 days. A liquid seal spray can be used as well as straw, sacking, etc., to cover cement concrete. Use heat, steam, canvas or plastic tarps, forms, sawdust, sand, and dirt as well as air entrainment to protect concrete from losing its strength.

Additives may be used at mixing time, or applied later after finishing cement concrete, in order to strengthen and protect it. Whether used for footings, foundations, columns, girders, beams, spandrels, lintels, floor slabs, decks or platforms, stairs, bases, or a complete structure, whether on or in the ground or water, concrete can be made more durable and be protected from the weather during its formation and curing by one or more various additives on the market. The manufacturer's directions should be followed carefully. Some additives not only lessen the amount of water required to produce good concrete, but also increase plasticity. They can be added directly to the mix. (Air-entrained portland cement contains such an agent in percentage quantities). Other additives enhance the bonding quality of a mix—good for patching and repairing. Hardener additives are used extensively to insure longer lasting concrete that withstands the elements and heavy traffic. Waterproofing additives come in such forms as crystal, powder, mastic, liquid, asphalt, oil, and pitch. Airtight seals and sprays seal in moisture.

Depending on the pour, finishing tools include brushes, hawk, screen, float, hammer, chisel (preferably 2″ wide), bucket trowel, bucket, Carborundum stone, and such other tools like power-operated chippers or bush hammers. The concrete is readied for filling all holes and grouting.

A filling mix of portland cement or white portland cement as a binder and sand like that used in the pour is made as follows: 1 part concrete, 1 part sand, and 1¼ or 1½ parts aggregate, mixed heavy with little water and allowed to stand before using. If the concrete wall is dry, wet it before filling holes. Fill in and float down; let it almost set before cutting or scraping level with the wall.

Grout is made of the same material, except the sand is very fine and the mix is rather thin, like cream. After honing, the wall is dampened. Grout is applied with a stiff brush to insure filling all small air holes. The grout is floated immediately after application. When the grout has taken up (not completely set, but set well enough not to pull out of the holes), it is scraped from the wall. A scraper or a trowel may be used to scrape off the semiset grout. Rub down the wall with sacking to remove loose surface grout after scraping, and again when the wall is sufficiently dry. No film should be left on the wall at the end of a day's work.

VOIDS IN MIXES

Wet sand volume bulks up as much as 26 percent. This is taken into consideration when proportioning by volume by increasing the sand and decreasing the water. Surface moisture in coarse aggregate is unimportant. The aggregate carries very little water unless it is porous as pumice or wet from damp or rainy weather. Weigh the sand, dry it out, and weigh it again to find the water content. Find the amount of buildup. Make allowances for volume quantity of both sand and water. In precise work this option may become necessary.

Good concrete mixes should be free of voids to attain compressive strength, durability, and imperviousness to water. To eliminate voids in proportion of materials, use a 1-cubic-foot container filled with dry aggregate and struck off level. Pour water from 1-cubic-foot container until the 1-cubic-foot container of aggregate is filled. Since a cubic foot of water weighs 62.4 pounds, the weight of the remaining water subtracted from 62.4 leaves the weight of water in the voids. From this, figure the void percentage. To find the voids in the aggregate and sand mix, again use water. Figure the cement necessary to fill the voids. Be certain that the cement encases all aggregate particles.

Because adding water to well-graded sand is difficult, another method can be used. Determine the percentage difference of voids for the solid to the loose particles of sand. Specific gravity of *siliceous* materials (limestone, sandstone, and quartz) is 2.65, which means 2.65 × 62.4 pounds or more than 165 pounds. The weight of a 1-cubic-foot container with dry, well-graded sand is 95 pounds. Therefore, (165 minus 95) ÷ 165 = 43 percent.

Cement concrete is a combination of cement, aggregate, and just enough water to produce the chemical action needed to bind the mix into a monolithic unit. Aggregate quantity should be no more than will be completely enveloped by the cement binder. The mix should be workable and of sufficient plastic consistency to let it be poured where required. A stiff mix will not pour into a narrow form or be worked easily when placed in a thin slab, but it works better in wide forms and deep pourings.

Excess water in a mix dilutes the cement binder and impairs the strength, durability, and watertightness of concrete. Thus, when mixing by hand, more cement is required. Machine mixing requires less cement. The same is true in compaction methods. Instead of hand spudding or ramming, the high-frequency concrete vibrator will compact and free the concrete of honeycombs or air bubbles. Less cement can be used. The ratio of fine to coarse aggregate can also be reduced, resulting in a stiffer and more satisfactory mix for finishing.

COMPOSITION FINISHES

Composition finishes are used to cover all kinds of masonry, concretes, and bases as protective and decorative cover coats. Some composition mixes require special treatment and are made with materials other than masonry mixes. Generally, the liquid agent is water, sometimes with or without additives.

TERRAZZO

Terrazzo masons can produce a beautiful, long lasting finish, whether this finish is applied to masonwork or to other material bases. Terrazo mixes require cements and aggregates in varied quantities. The cements can be portland gray cements or other kinds. White cements are also used. The aggregates include chips of marble, marble dust, gemstone, bluestone, mother-of-pearl, crockery, glass, and enamel.

Mixes

The terrazzo underbed mix is 1 part portland cement (binder), 4 parts sharp graded sand (aggregate), and enough water to give the mix a plastic state. The terrazzo finish mix is 1 part cement, gray or white (binder); 2 parts aggregate chips, size from 1/10″ to 1/4″; water, enough for a plastic, not too wet mix; metallic color, if desired; and calcium chloride, from 1 pound to not over 4 pounds per 1 cubic foot of binder. Flakes are added to the aggregate.

When terrazzo is used as a finish over concrete pours, the concrete base should be 2″ thick below the finish surface. The aggregate for the underbed is 1 part sand, and 5½ parts hard coal cinders, or other aggregate such as gravel, pumice, slag, or broken stone. Put the aggregate into the machine or mortar box. Add the binder. Mix well. Add the water—just enough to bring the mix to a plastic consistency for pouring. After pouring it should be screeded level, leaving at least ½″ but not over ¾″ below the finish surface for the terrazzo.

On this underbed before it is set, brass strips as parting strips or grounds are installed. Generally they are placed 4″ apart. They can be straight, curved, or any shape desired. Also on the underbed, the various decorative units are placed and secured with a grout mix (cement and water, "neat grout"). The decorative units and brass or other material grounds should come to the finish surface.

This underbed, as well as the base the underbed is put on, should be cleaned of all rubbish, chips, plaster, oils, etc. Do not wash these surfaces with an acid solution. All concrete or masonry bases should be wetted well to insure cohesion.

The terrazzo mix requires the binder and aggregate to be mixed well before the water is added. The aggregate should be graded from fine to ¼″ size particles, and all should pass through a ⅜″ screen. The binder should be free of lumps and all set material. The water should be free of all sulfur and other impurities. An additive is not needed except for hardness and quick-setting purposes. A cover wash other than a grout is not required unless needed to seal in moisture to insure even setting.

The terrazzo material should be poured in the spaces between the brass dividing strips. The terrazzo can be leveled off and should be rolled and packed with a smooth roller to free it of superfluous water. It is then troweled to an even true surface. The troweling should bring the material level with the brass strips. The troweling is best done by hand, although troweling machines are used.

The method of terrazzo work is different than on bases which will not crack. The procedure of terrazzo work on these bases is as follows. The mixing proportions are the same for the underbed and terrazzo finish as for the masonry base. The structural base should be at least 3″ below the finish terrazzo surface. The base is covered with a level bed of dry sand. Tarpaper, 30 pound felt, and plastic are placed on this sand. The purpose is not to disturb the sand and to keep the underbed separated from the sand.

The underbed should be at least 2″ thick and should come up to ½″ to ¾″ of the surface of the terrazzo finish. The brass strips and desired decorative units are secured on the underbed. They should be level to the finish surface of the terrazzo. The terrazzo is run on the underbed as was in the terrazzo work on the noncracking base.

When terrazzo work is laid on wood or other bases, the procedure is different in some respects. The thickness of the terrazzo base should not be less than 2″. The base should be clean and dry. Tarpaper and 15-pound or 30-pound felt are laid on the floor. Galvanized wire, of at least 14 gauge 2″ mesh, is nailed down on the paper. The underbed is leveled off on this paper and wired to a depth which leaves at least ½″ to ¾″ space to the brass strips grade. The strips are placed in the underbed. The method of laying the terrazzo is the same as on the other bases.

When a terrazzo baseboard or wainscot is laid, the brass strips are generally placed 4′ apart. A base bead should be nailed true to level as a ground for the terrazzo work. The underbed should be run on a base which has metal lath as a hang cleating surface. The underbed should be of a thickness which leaves at least ⅜″ room for a ⅜″-thick coat of terrazzo. The terrazzo, as well as the undercoat, is placed by a plasterer or cement mason and done with a trowel.

Terrazzo partitions are also carried out the same as wainscot terrazzo work. Galvanized metal expanded lath and metal studding should be used in the partitions. The partition should have a scratch coat (of 1 part cement and 2 parts sand mix, with enough water to bring it to a plastic state). The scratch coat should be at least ¼″ thick. It should be plumb, flat, and true, and brought to within ½″ to 1″ of the terrazzo finish surface. A second coat is run for trueing to ½″ to ¾″ to the finish surface. The underbed should be run to within ½″ to ¾″ of the finish surface. The total thickness of this work should not be over 2¼″.

All terrazzo work after troweling should be sanded with a Carborundum stone or similar abrasive stone, either by hand or machine. The terrazzo on the walls, base, and wainscot should be honed to a fine finish.

After sanding, the terrazzo should have a coat of pourable grout (neat). This grout should fill all voids. It, too, is sanded and thoroughly washed clean.

Magnesite Terrazzo

Magnesite terrazzo composition material is somewhat different than the cement terrazzo composition material. The materials necessary to make magnesite terrazzo are calcined magnesite aggregate, magnesium chloride solution, and color. The binder is magnesite (magnesia). Magnesite is made from chalklike magnesium limestone by calcining. This calcined powdered magnesite should be kept dry and in tight drums.

Aggregates used can be clean white graded sand, silica, powdered quartz, soapstone, ground cork, marble dust, asbestos, talc, pumice, feldspar, wood pulp, and sawdust. Fabricated aggregates which can represent different natural aggregates may be used. All wood powders, sawdusts, and wood chips should be treated for liquid proofing as they may swell and cause cracking. They can be dipped into a solution of glue and water and then drained well before using.

The liquid activator (a catalyst) is a magnesium chloride solution instead of water as in cement terrazzo. The solid crystal magnesium chloride resembles ice and is in solid chunks. It has to be broken up into small particles with a hammer. They dissolve in water and have to be dissolved in enough water to bring the magnesium chloride solution to a density of 14 to 28 degrees, depending on weather conditions. When the weather is cold (not freezing), the density on the Baumé hydrometer should be read from 22 to 25 degrees or even to 28 degrees. Water should be added to the magnesium the day before use as the crystals take some time to dissolve.

Use dry paint colors. Do not use chrome colors as they destroy the compound chemically. Colors like black and gray should be from 1 percent to 4 percent, according to the shade desired, by weight of the magnesite binder. Colors like blue, green, and yellow should be about 8 percent of the weight of the magnesite binder. Colors like red, brown, and buff should be about 10 percent of the weight of the magnesite binder.

The necessary tools and equipment are a Baumé hydrometer, shovel, hoe, scales for weighing, mixing box, screen and flour sieve, two 10-quart galvanized pails, trowels, brass strips, galvanized metal lath, galvanized 14 gauge 2″ mesh welded wire, nails, roller, screeds, and power sanders.

All grounds should be true and level to the finished surface. They should be securely installed along with the decorative units.

The magnesite varies in setting time in proportion to the percentage of carbon dioxide it contains. Therefore, it is best to use one manufacturer's magnesite throughout the job. The binder is magnesite plus magnesium chloride (sorel cement). The aggregate can be natural, artificial, fiber, or a combination of these in graded or powdered form. The liquid agent is water. If the aggregate is graded, maximum size is not over ¼″.

The formula is 1 part magnesite, 2½ parts graded aggregate or 4/5 part powdered aggregate (by weight), and enough water to bring the mix to a plastic and not too stiff state.

Thorough mixing is essential in all masonry mixes to produce an even textured and durable product. Improper mixing is the cause of failure, providing the temperature is stable throughout the setting and curing stage. If the temperature drops more than 15 degrees, the pour cracks. If it rises more than 15 degrees, the pour expands and bulges up. After 6 to 12 days, the pour is safe, durable, beautiful, and almost indestructible.

Cement concrete or cement plaster, thoroughly cleaned and treated with a terrazzo grout, is the best underbed for terrazzo. A good mix of 1 part magnesite, ½ part very fine sand, and enough magnesium chloride water (20 degrees in density) makes a thin grout. All bases, including underbed, should be at least six days old. Wood bases should be varnished with asphalt and nailed well. Only galvanized wire should be used.

On solid walls, a ½" deep scratch coat is applied and left rough. When it is thoroughly dry, give it a coat of thin grout. Follow with the finish coat, screed to grade, darby true, and then let it take up. When taken up firmly, troweling can be done. Trowel several times during the initial setting. Care should be taken in respect to blisters. Uneven troweling, when it has hardened, makes a mottled surface, especially of the grays. Well-troweled surfaces take a better polish. This material can be finished as is terrazzo or, when semihard, it can be ground clear and smooth with a pumice block and water. After it is thoroughly dry, it can be coated with wax or rubbed linseed oil and polished to a shine with a wool cloth.

Formula Mixes

Mixes for panels, tiles, and concrete faces can be mixed thinner and poured to a depth of more than ¾", as for floors and walls, or to a depth of 2". Pour on plate glass that is oiled with olive oil or neat's-foot oil. Use clamps to fasten on the grounds. Set in stone as soon as the mix is poured and let set until thoroughly dry. Molded forms may be used for rustic work. Here are some formula mixes: 1 part magnesite, 1/5 part asbestos fiber, ¼ part silica, 1/10 part chips, and 1/10 part brown color; 1 part magnesite, ¼ part asbestos fiber, 1/5 part silica, 1/5 part talc, and 1/10 part red color; 1 part magnesite, ¼ part asbestos fiber, 1/5 part talc, ¼ part silica, and 1/10 part black color; and 1 part magnesite, ¼ part asbestos fiber, ¼ part chips, 1/5 part talc, and 1/10 part yellow color. For imitation marble, use 1 part magnesite, ¼ part silica, 1/6 part marble dust, ¼ part terra alba, and ¼ part talc.

Magnesium chloride and water of 26 to 28 degrees density is the liquid agent for these formula mixes. For best results, never make these mixes or pour them if the temperature is below 55° degrees or higher than 60 degrees Fahrenheit.

The following two formulas can be lined with colors, as in unit making, to represent marble. The liquid agent is water: 1 part Keene's cement, 1/5 part marble dust, 1/5 part talc, and 2/5 part silica; and 1 part portland cement, ¼ part Keene's cement, ⅜ part silica (fine grade), and 1/12 part hydrated lime. These mixes are handled with the same care as terrazzo when chips are used in the aggregate.

Metal strips can be used for this work. It is not as necessary to watch the temperature closely as when the binder is magnesite.

Chapter 7

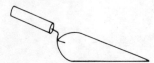

Facets of Masonwork

The first thing to consider when deciding to build is location, which should be a place free from landslides, snowslides, and swamps. The location must be where there is good drainage or where a drainage ramp can be placed around the building. The location should be in a place where this kind of construction is allowed.

The method of *layout* is considered next. The *plumb bob* locates the marginal point of the corner of construction (Fig. 7-1).

A good footing and foundation are important. The concrete floor of the basement should be reinforced to stop cracking. All concrete work should be reinforced. The units of construction stay secure much longer when reinforcing is used. Sidewalks should be considered (Fig. 7-2).

After the foundation is ready for the house, you must choose construction units. This material should be easily available for use.

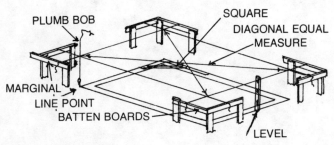

Fig. 7-1. A layout method.

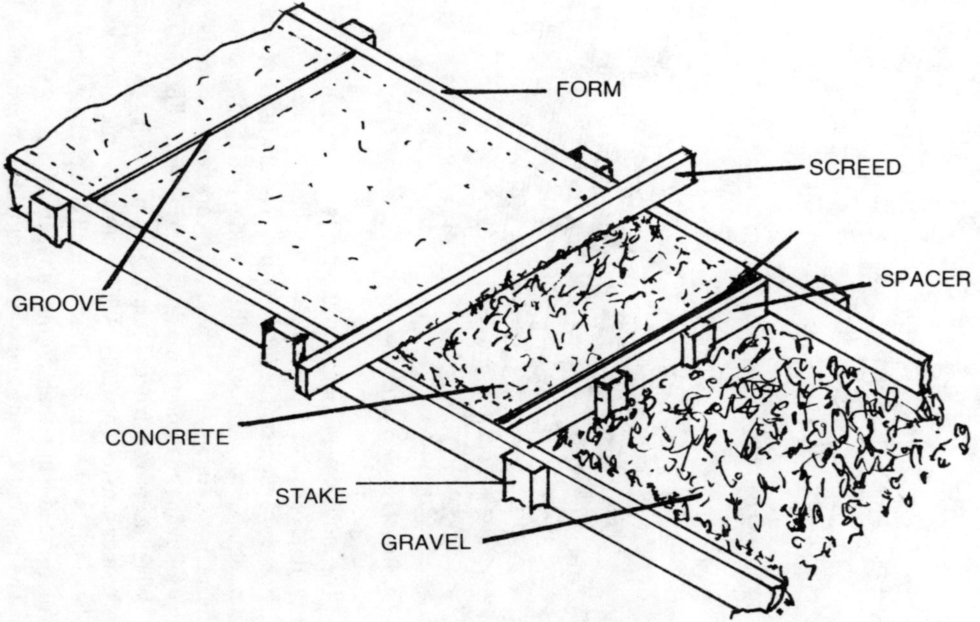

Fig. 7-2. Concrete walkway.

FORM
SCREED
SPACER
GROOVE
CONCRETE
STAKE
GRAVEL

The material includes units, materials for mortars, scaffolds, etc., that are used in the masonwork.

TROWEL HANDLING AND LAYOUT METHODS

The method of using the trowel on brickwork is shown in Fig. 7-3. The method of mortaring the mortar bed and end units are shown in the following stonework (Fig. 7-4).

Fig. 7-3. Trowel handling on brickwork.

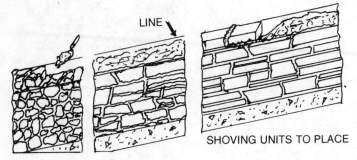

SHOVING UNITS TO PLACE

Fig. 7-4. Trowel handling on stonework.

The method of placing mortar on tilework is shown in Fig. 7-5. The method of using a trowel when laying block is shown in Fig. 4-9.

Figure 7-6 depicts the laying of the masonry units of construction in combination with each other. Stucco is also shown. A combination of brick, tile, plaster, asphalt, and cork construction is shown in the following method of making a refrigerator (Fig. 7-7).

When using construction units, it is necessary to lay out the units or measure for openings. A "give and take" method is used for the head joints of some construction units, especially brick which has more head joints than do the other units (Fig. 7-8).

The rise of the units and courses should be measured so they, when laid, come out even with the top of all openings as well as at

Fig. 7-5. Trowel handling on tile.

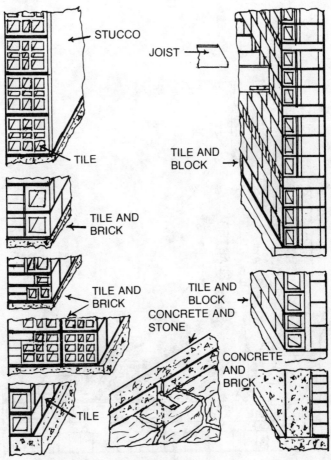

Fig. 7-6. Examples of mixed unit construction.

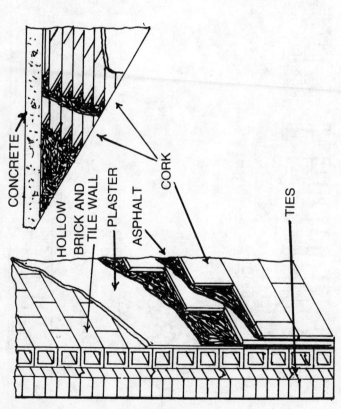

CONCRETE

HOLLOW
BRICK AND
TILE WALL

PLASTER

ASPHALT

CORK

CORK

TIES

Fig. 7-7. A method for making a refrigerator.

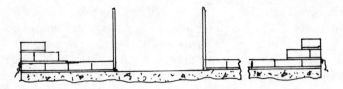

Fig. 7-8. Spacing detail.

the top where required. Both ends of courses should rise the same, or the course comes out even at one end and not at the other. When laying masonwork, measure from that which one is to lay to the course just laid.

LAYING TILE

There was a time when natural tile had no underbed. The tiles were placed in a haphazard manner and eventually seated with windblown dust and gravitation. Tile setting has since become an art. The methods of tile setting have evolved through many constructional periods. Many of these methods are still used.

Most tile laid today requires an underbed. An underbed may be of various materials, compacted or set, whichever is suitable for the job.

Method One

Some outside patios and walks have underbeds of compacted gravel and sand or clay mix. The depths should be the same and have slopes for drainage. A dry mix, consisting of 1 part portland cement to 3½ parts sand, is sprinkled over the area to a depth of 1″. Natural tiles are laid to grade on this bed. The joints are tamped full with this same mix. Fine sand is spread over the work to a depth of from 1″ to 2″. The surface is sprayed with water until all of the mix is saturated. The sand is removed by the third or fourth day. All joints are checked, and the tiles are cleaned.

Method Two

The compacted gravel underbed may be covered with a 1″ layer of good soil and the tile laid to grade, with the joints filled with the soil. Grass can be planted in the joints. This patio or walk needs watering until the grass is up. This kind of work withstands frost. Anyone can do it.

A more solid and long lasting job is done as follows. Set the forms and grounds to the required grade for the underbed; 2″ × 4″

wood grounds and forms may be used as the depth of the bed need not be over 4". Tamp in the gravel and sand mix. Wet it down before adding to the height of the forms. Pull up any cross grounds you may have used. Fill in the space they occupied and tamp well. Place 9 gauge, 6" square-mesh welded reinforcing over the underbed. Add forms, at least 3", and bring them to grade. Soak the underbed with water. The purpose of wetting down the underbed is to keep it from absorbing the moisture from the cement mix.

The underbed mix is made of 1 part portland cement, 3 parts graded sand, and 2 parts graded aggregate. Use just enough water in the mix to make it plastic. Pour in the mix. This mix should be leveled off at a grade approximately three-fourths the average thickness of the tile from the finished grade. If the average thickness of the tile is 1⅜", the mix should be brought to within ⅞" from the top of the forms. The poured mix plus the tile should never be less in thickness than 3½". The tile should never be thicker than one-half the thickness of the mix plus the tile.

Pour only as much mix at a time as can be tiled before the initial setting of the mix. All excess mix coming through the joints when the tile is tamped down to grade should be picked up with a trowel and placed in the poured mix only if it has not taken its initial set. All joints should be filled with mortar. As soon as the laid tile can be walked on, all the joints should be brushed crossways with either a stiff scrub brush or a steel brush. The tile should be cleaned with the use of sharp sand.

Tile Thickness

Tile setting by tilemasons is performed differently than the methods just explained. Their methods are more in line of perfection, exactness, and are of a more artistic nature regarding appearance, durability, and soundness.

In order to have the monolithic pour, base foundation, and slope at the correct grade to accommodate the required underbed, float coat, and tile, it is necessary to know the thickness of the tile to be used. This is abolutely necessary so the finished surface of the tile comes exactly to the intended floor grade and required face surface of a floor and wall.

Reinforcing

When the thickness of an underbed is 3½" or more, it is not necessary to use reinforcing steel or welded mesh. If the bed is from 2" to 3½" thick, it is best to use welded wire or reinforcing steel.

The reinforcing steel bars should be run at right angles to each other. The welded wire squares should not be over 10″ and need not be smaller than 4″. Should the bed be less than 2″ thick, 4″ or 6″ welded mesh is used as reinforcing. Reinforcing keeps the bed from cracking if a good concrete mix is used for the underbed.

Float Coat Application

The float coat is generally from ½″ to 1″ thick. This coat is generally laid ⅛″ higher than the bottom of the tile used. This extra ⅛″ is for seating the tile to place. When the practice of buttering the tiles is used, the float coat is brought to a grade which will allow just the thickness of the tile from the finished floor grade. On walls the float coat is seldom more than ½″ thick. This coat is also ⅛″ more in thickness than is required when the seating method of tile setting is practiced. These two methods of applying the float coat are done with a plaster's trowel and screeded perfectly to grade. The float coat should never set up before the tile is placed; so it is best to not apply any more than can be tiled to grade before setting takes place.

When a tile floor is to be installed on a monolith, masonry, cement plaster, or wood base, the base shall be cared for as when a *terrazzo* floor is installed. The wood base shall be covered with an asphalt emulsion or tarpaper sealed to keep moisture from swelling the wood. The underbed can be placed on any of these bases. On a wood floor it may be necessary to use a much thinner underbed, or it may be necessary to dispense with the underbed. The floating coat may be a mastic applied to the welded mesh directly, to a depth of not less than ½″. It is on this mastic that the tiles are laid.

After the underbed has been poured and screeded to the required grade, and left in a true rough finish, the float coat can be applied. It is not necessary to apply a reinforcing mesh if the underbed is reinforced. Grounds should be held in place with cement. Have the grounds placed or seated in the underbed, just to a depth necessary to hold them. They can be placed to the finish floor grade before the underbed is set and removed as the tile laying progresses. Only as much float should be laid as can be tiled before the float coat takes up. It should lose its shine before the tiles are placed, or the tiles will sink in and be below grade. Dry cement can be sifted or thrown on the float coat if necessary to take up the moisture.

If the float coat has been screeded to a true grade, on the floor or wall, the tile should be buttered quickly, with some heavy cement grout on the bottom of the tile before it is placed. If the float coat has been screeded to the thickness of the tile from grade, the tile can be dipped in a thin cement grout and laid. In both methods, tile must be placed straight with the work and have the required joint spacing. When the float is covered with tile and taken up some, the tiles are ready to be seated to the required depth. A block of wood and a heavy hammer are used to pound them to grade. A straightedge is placed from ground to ground for the purpose of checking the tile. A good practice is the use of dry mix of 1 part cement to 1½ to 2 parts sand. Place this in the joints. This dry mix takes up at the bottom of the joint and holds the tile from shifting during the leveling process.

When the tiles are seated to grade, the dry mix is cleaned from the joints. The joints are now filled with the joint mix. The grout mix is poured on the tile and worked into the joints with a flat trowel. When the joint mix has taken up, the tiles are cleaned. Sawdust is used to thoroughly gather up the grout mix from the tile. When the tiles are set well, they may be rubbed with fine sand if any mix is still on them. Soap and water may be used to clean the tile. If muriatic acid is used, the solution should never be stronger than 1 part acid to 12 parts water. See Figs. 7-9 and 7-10.

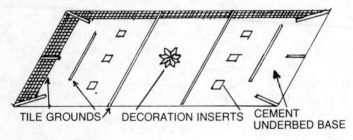

TILE GROUNDS   DECORATION INSERTS   CEMENT UNDERBED BASE

Fig. 7-9. Tile floor layout.

Walls and Floors

Walls are laid in the same manner. All base coats on walls where tiles are to be laid should be of cement mortar. The base coat should be brought to an even flat surface and left rough to receive the float coat. The float coat should be screeded to a perfect flat surface.

There are a few pertinent things to remember in all masonwork. Never apply any mix on a totally dry surface. It will not unite with the masonry surface. Never apply any mix on a moisture-logged surface. It will not set up quickly. Never apply any mix that is subject to water after becoming hard to any surface not sealed with a waterproof material, such as portland cement. Never apply any mix

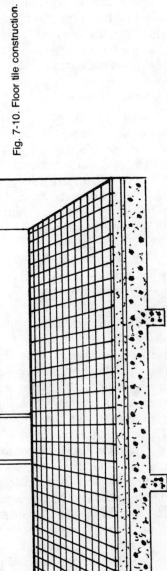

Fig. 7-10. Floor tile construction.

which can be affected by the chemicals in the mix to which it is applied.

Vitreous and semivitreous tiles are the best tiles for floors. Abrasive tiles rather than slick glazed tiles are safer tiles for floors. Always protect finished work until it is safely set and in condition to use. Never allow any masonwork to freeze or dry out before it is thoroughly set and cured.

Spot Method

When laying tile, the *spot method* rather than the ground method is used by many masons. This is accomplished by spotting tile 4' to 6' apart. They should be spotted parallel or perpendicular to the laying. Take a tile, butter the back, and press it to the wall or floor underbed. Use a straightedge or screed, and possibly a level, to align it to grade or the finished face of the work. Several tiles spotted to grade serve the purpose of grounds. They can be removed when the work progresses to them.

When laying out the work, find the center of the wall or floor (Fig. 7-11). All tilework, on any job, should be started from the center and progress outward, especially if there are any patterns or designs in the work. Diagonal lines can be stretched to find the center.

Any decorative panels or figured design units should be properly placed. The tiles should then be figured and laid. Sheet tiles are handled much the same as single tiles. Care must be taken that the joints between the sheets are in comparison to the joint sizes of the joints in the sheet tile. The paper backing should be left on until the tiles are seated and set enough so they will not be moved when the paper is saturated with water and removed.

The general size of joints are as follows: 1/16", 3/16", ¼", ½", and ¾". They are listed in order as to the size of tile units. The larger the unit, the larger the joint size can be. Joint sizes of ¼" (sometimes), ½", and ¾" are tooled with a jointer. Tooling should not be done until the mortar is semiset.

The underbed mix to use is 1 part portland cement, 2½ parts sand, 2 parts one-half maximum aggregate, and water. This mix is for a 2" or more thick underbed. A mix of 1 part cement, 2 parts sand, 1 aggregate of one-half maximum size, and water is for an underbed of 2" or less in thickness. No underbed should be less than 1" thick. The underbed on walls may vary in thickness when the wall is of masonry. Its plaster coat should be at least ½" thick and of a 1 part portland cement, 2 parts sand mix. It should be left rough

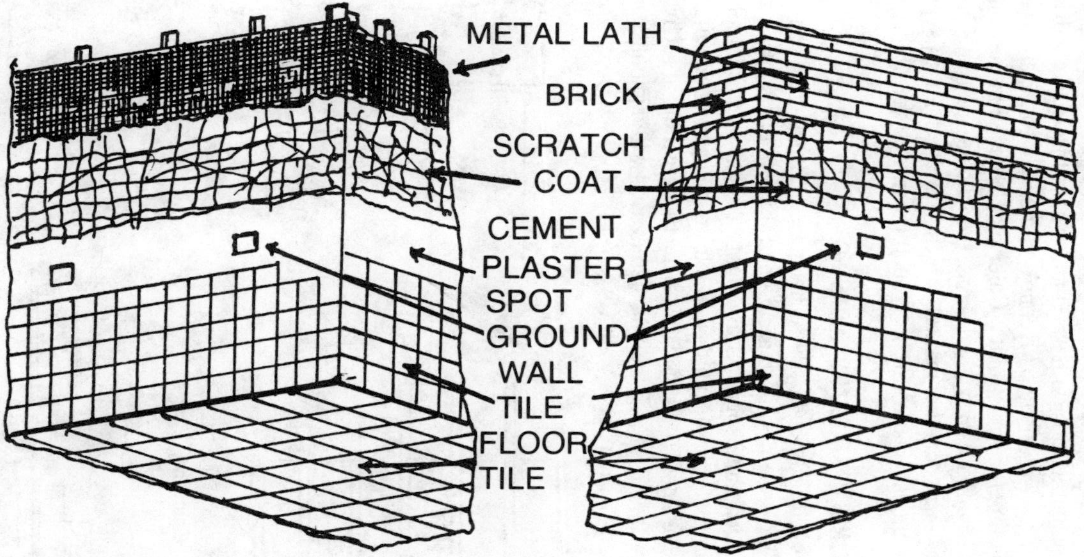

METAL LATH
BRICK
SCRATCH COAT
CEMENT
PLASTER
SPOT
GROUND
WALL TILE
FLOOR TILE

Fig. 7-11. Floor and wall tile layout.

137

scratched to receive the float coat. The float coat should be a mix of 1 part portland cement, 1½ to 1¾ parts sand, and water.

The joint mix can consist of portland or Keene's cement and can contain 5 percent hydrated lime in the binder. This is a mix in a stiff or semistiff form with water and is used for small width joints. A mix of 1 part portland or Keene's cement, 1 to 2 parts sand, and water is fine for wider joints. Color may be used in these mixes.

It is unnecessary to discuss the cutting of tile. There are *diamond saws* on the market which are used for this purpose. Uneven unglazed tile can be ground down like terrazzo.

CORBELING, RAKING, AND TOOTHING

Corbeling is used where necessary in masonry construction (Fig. 7-12). A corbel is an architectural member that projects from within a wall and supports a weight. *Raking* back is employed at leads and where the units are left to be continued (Fig. 7-13). *Toothing* is another method of leaving the laid units in such a way as to be continued later. Toothing also is used when joining partitions to other partitions or a walk (Fig. 7-14).

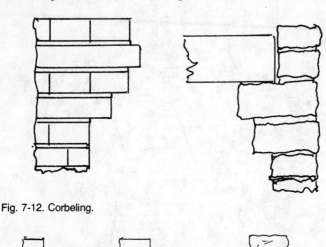

Fig. 7-12. Corbeling.

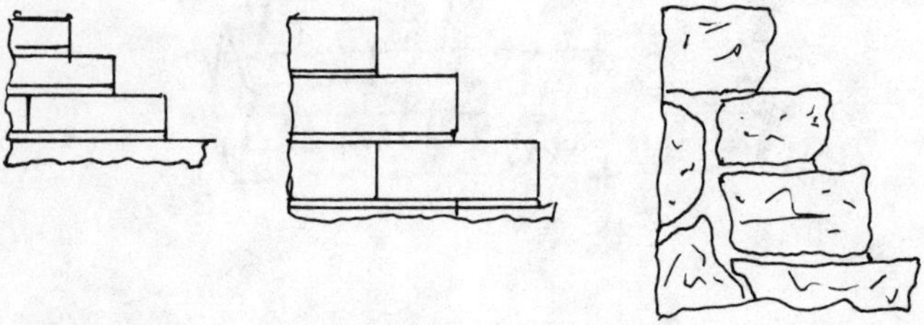

Fig. 7-13. Raking.

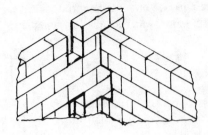

Fig. 7-14. Toothing.

REINFORCING ITEMS

A *bolt anchor,* which is used to fasten the plate, is pictured in Fig. 7-15. Dur-o-wall and re-bar are also used. Re-bar is ideal when reinforcing perpendicularly.

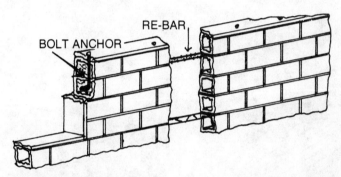

Fig. 7-15. Reinforcing items.

The electrical conduit can be run horizontally as well as perpendicularly through masonwork. All reinforcing, ties, etc., should be bedded in. The electrical conduit and water or steam pipes do not have to be bedded in mortar.

Some masonwork is laid up with an air space. This space uses ties to hold the two walls together. This hollow space may condense, so *weep holes* are installed at the wall bottom, or some of the first course head joints are left out. This lets water condensation out of the wall. Insulation can occupy this hollow space when required.

The outer wall of a hollow wall is often plastered as it is laid up. Generally it is laid first, then plastered before the inside load-bearing wall is laid. I have never laid the outer wall over 2' high before I plastered and laid up the inner wall.

Figure 7-16 shows the *pier,* the *pillar,* and the *pilaster.* When a pilaster is not used, a metal post is used within or inside the wall for the purpose of containing the load. The masonwork between rafters, joints, etc., is called nogging (Fig. 7-17).

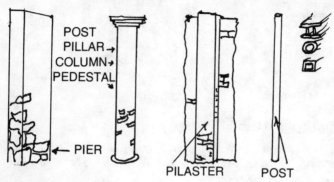

POST
PILLAR→
COLUMN→
PEDESTAL

← PIER

PILASTER   POST

Fig. 7-16. Bearing parts.

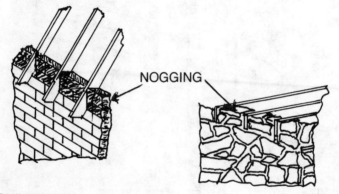

NOGGING

Fig. 7-17. Nogging examples.

LAYING WINDOW SILLS AND GLAZED TILES

The window sills should be laid with care. A line should be used when two or more sills are in alignment to keep them horizontal with each other. The sill should have drainage out, not in. There should also be a groove under the sill lip, if possible, which keeps water from running back to the wall (Fig. 7-18).

Glazed tiles are made in various sizes and are used in places that have moisture like bathrooms and kitchens. The mortar used should be water-resistant. This method of laying glazed tile is shown in Fig. 7-19.

There are inside and outside wall tiles, glazed and unglazed. These tiles make good, long lasting, practical and decorative covers. Figure 7-20 depicts two walls using tile. Tile needs a cement plaster base.

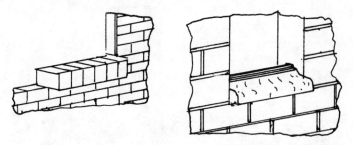

Fig. 7-18. Lay window sills with care.

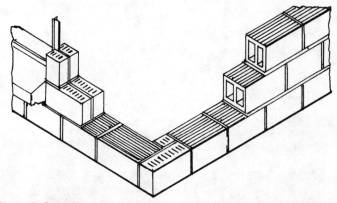

Fig. 7-19. Glaze tilework.

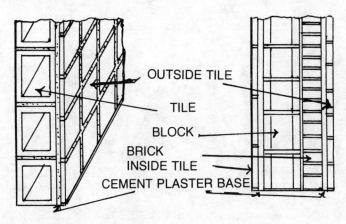

OUTSIDE TILE

TILE

BLOCK

BRICK
INSIDE TILE
CEMENT PLASTER BASE

Fig. 7-20. Two walls requiring tile.

STONEWORK

Figures 7-21 through 7-30 depict stonework. Figure 7-29 shows the scaffolding used when setting rubble stonework.

Fig. 7-21. Stone garden wall.

Fig. 7-22. Wrought ashlar stone.

Fig. 7-23. Stucco and cobblestone.

Fig. 7-24. Cobblestone.

143

Fig. 7-25. Springer stone arches.

Fig. 7-26. Note the posts.

Fig. 7-27. Ashlar bond stonework.

Fig. 7-28. Ashlar set stonework.

Fig. 7-29. Scaffolding used when setting rubble stonework.

Fig. 7-30. A stone house.

146

Chapter 8

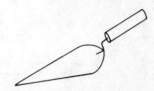

Lintels and Arches

The purpose of a *lintel* is to hold up the weight load above an opening. All lintels should be of a breadth equal to, or more than, the thickness of the wall. Lintels can be of masonry units, metal, concrete, or wood. A lintel should extend at least 4″ into the wall at its ends.

LINTELS

Lintels span openings and support weight. A lintel's span opening should be no wider than the lintel is capable of sustaining the load weight. A good lintel is shown in Fig. 8-1.

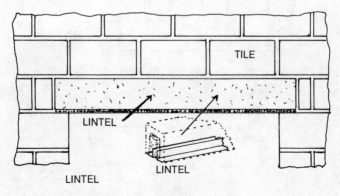

Fig. 8-1. A lintel is designed to support weight.

A concrete lintel should be reinforced as in Fig. 8-2. A concrete-filled and reinforced tile lintel is shown in Fig. 8-3. A concrete-filled and reinforced block lintel is shown in Fig. 8-4. A concrete-filled and reinforced 16″ high beam block lintel is shown in Fig. 8-5. A metal lintel is shown in Fig. 8-6, and a wood lintel is depicted in Fig. 8-7.

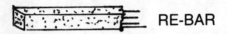

Fig. 8-2. A concrete lintel.

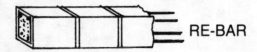

Fig. 8-3. A concrete-filled tile lintel.

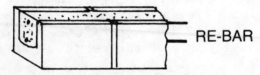

Fig. 8-4. A concrete-filled block lintel.

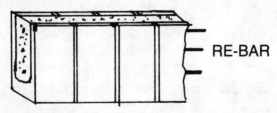

Fig. 8-5. A concrete-filled and reinforced 16″ high beam block lintel.

Fig. 8-6. A metal lintel.

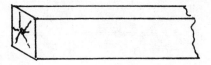

Fig. 8-7. A wood lintel.

ARCHES

The purpose of an *arch* is to span openings and support weight above them by resolving vertical pressure into diagonal or horizontal pressure. For this purpose, its breadth at least equals the thickness of the wall.

Straight Arch

A *straight* or *flat* arch takes the place of a lintel. Because this arch is weak, it is necessary to weld together angle irons and use them to strengthen it (Fig. 8-8). The *extrados* (upper, outward curve of an arch) may be slightly curved or straight. This arch, if straight, appears to sag in the middle. A slight convexity of the intrados corrects this (the extrados rising in the same curvature).

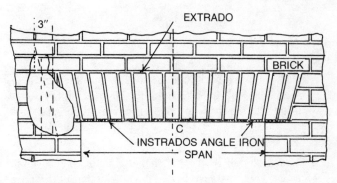

Fig. 8-8. A straight arch.

Camber Arch

The *camber* arch and an angle iron strengthener are shown in Fig. 8-9. This arch can have its extrados straight or curved.

Structural Arch

The *structural* arch is often used in construction. This arch has a curvature and is useful when the span is not long (Fig. 8-10).

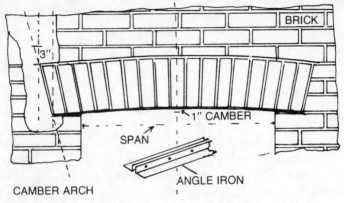

Fig. 8-9. A camber arch.

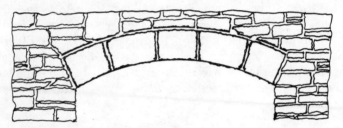

Fig. 8-10. A structural arch.

Buck

Arches require a buck to hold them up until they are set (Fig. 8-11). Cross slats are used to hold up the voussoirs. The spaces between them leave room to rake out or tool the joints.

Segmental Arch

A *segmental* arch is stronger than the flat or chamber arch. Its perpendicular rise (distance between the key unit of the arch and the spring line) is greater than the straight arch, so it resolves more vertical pressure into diagonal thrust. This arch does not need angle irons to strengthen it. The rise of a segmental arch is generally less than one-fourth and more than one-twelfth the width of the opening. The lower the rise of a segmental arch, the weaker the arch is. Therefore, the abutment at the skewback has to be stronger in order to resist thrust.

The method of setting out the segmental arch is shown in Fig. 8-12. Strike the rise line on the layout board. Strike the horizontal

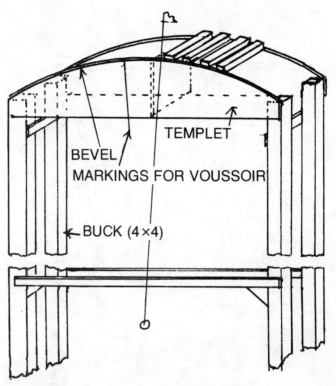

TEMPLET

BEVEL

MARKINGS FOR VOUSSOIR

← BUCK (4×4)

Fig. 8-11. A buck is used to hold up arches until they are set.

line (spring line) at right angles to and across the rise line. Where the lines cross is the center of the arch and the center of the span opening designated by C. From point C along the spring line to point 2 is a distance equal to one-half the span of the opening. The rise of the arch is the distance from the spring line at point C to the bottom of the arch key or center unit (from C to 1). The rise of the arch should have a height of one-half to one-fourth the span's opening along the spring line. With a compass point at point 2, open it to point 1. Scribe an arc from point 2 and point 1, above and below the spring line. Do the same with the same distance in the compass from point 1. Arcs intersect at points 6 and 7. By striking a line from point 7 through point 6 and cutting through the rise line, the radial point 8 is found. The compass point is placed at point 8, opened up to point 2, and scribed to point 3 for the intrados of half the arch. The extrados is scribed by extending the compass the breadth of the arch face, which should not be less than the thickness of the wall, to

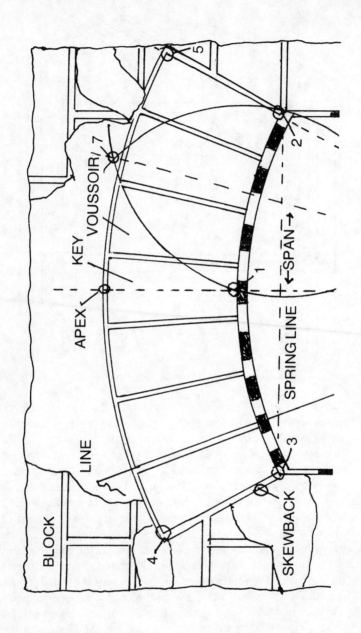

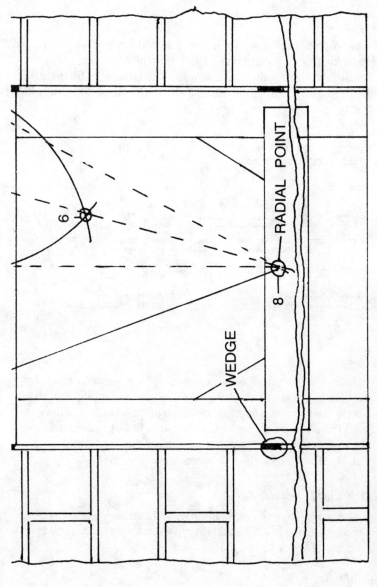

Fig. 8-12. A segmental arch.

153

points 4 and 5. The inclination of the units and skewback is in alignment with the radial point 8. The line comes in handy when laying the units of the arch (Fig. 8-12).

Relieving Arch

A *relieving* arch is generally placed inside masonwork to relieve vertical pressure and distribute it from places of weakness in the wall. It is also used above openings. The tympanum filler core lacks strength as a structural member. It sets between the intrados of the relieving arch and a lintel (Fig. 8-13).

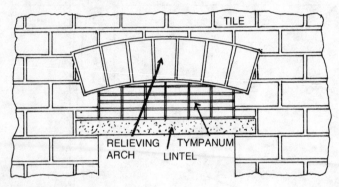

Fig. 8-13. A relieving arch.

Bull's-Eye

Bull's-eye is scribed with the compass from its center radial point. The key units at the top and sides should be placed with respect to the horizontal spring line and the perpendicular rise line.

All the units are aligned from the radial line at the center C. The units are sometimes larger and can be of different material to give the bull's-eye a distinctive appearance (Fig. 8-14).

Semicircular Arch

A *semicircular* arch has its intrados and extrados run from the center of the spring line. The units are in alignment with this center radial point (Figs. 8-15 and 8-16).

Gothic Arch

A *Gothic* arch springs from a horizontal surface at the spring line and is a very strong arch. The distance from point 2 on the

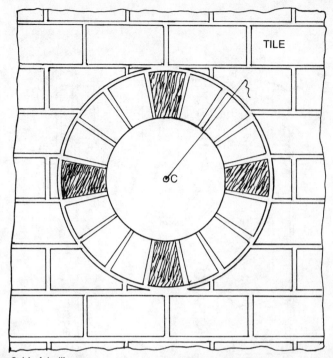

Fig. 8-14. A bull's-eye.

spring line to point 3, the distance from point 3 to point 1, and the distance from point 1 to point 2 are equal, forming an equilateral triangle. To scribe the intrados and extrados of the arch, the com-

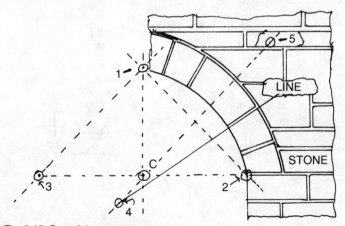

Fig. 8-15. Part of the semicircular arch.

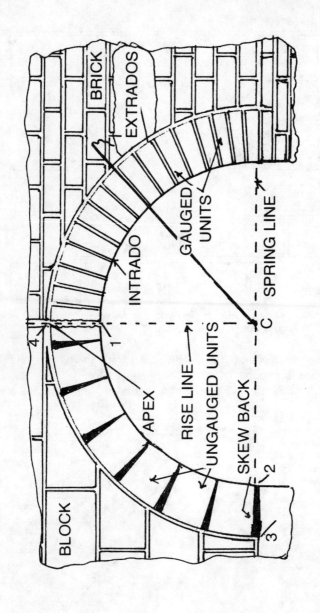

Fig. 8-16. The semicircular arch.

pass point is placed at points 3 and 2, respectively. The units making up the arch between points 2 and 1 are in alignment with spring line radial point 3. The units between points 3 and 1 are in alignment with radial point 2. The voussoir units are generally the same size. Should this arch have a bird's-mouth unit key, the unit alignment of the voussoirs has to be arrived at in a similar manner as in an elliptical arch (Fig. 8-17).

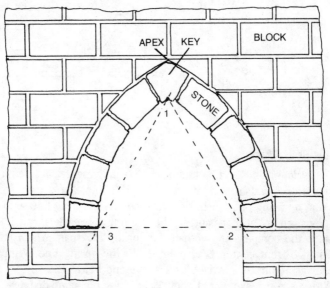

Fig. 8-17. A Gothic arch.

Low-Rise Gothic Arch

A *low-rise Gothic* arch is two segmental arches leaned together at the apex and is laid out like a segmental arch. The unit members are in alignment with the radial points 7 and 8 (Fig. 8-18).

Semi-Gothic Arch

A semi-Gothic arch is a beautiful arch. It is a strong arch. Its intrados, a semicircle, is scribed from radial point C. After deciding the width of the breadth face of the arch, points 3 and 4 are found. The breadth or face height of the arch from the rise point C to the apex point 1 may be as desired. The extrados can then be scribed by using the Gothic arch method. Unit alignment is from radial point C or from points 8 and 7 respectively, in which case the springer and skewback alignment are found. (Figs. 8-19 and 8-20).

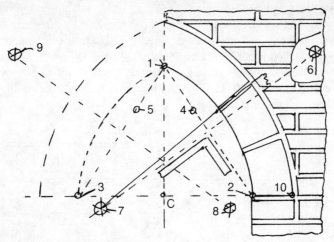

Fig. 8-18. A low-rise Gothic arch.

Low-Crown Arch

A *low-crown* arch is laid out by striking a vertical line on the layout deck and crossing it with the spring line. The arch height is found by using one-fourth the distance of the spring line with the compass point at C and cutting the rise line above the spring line 1. Open the compass to one-half the spring line span, which is the distance from center C to point 2, on the spring line. With the compass point at 2, scribe arcs above and below the spring line. With the compass point at point 1 and using the same distance, cut the two arcs just made to find points 5 and 6 at the arcs. Strike a line through points 5 and 6 and across the vertical center line of the arch, thus locating the radial point 7. Place the compass point at point 7, extend the compass to point 2, and scribe the intrados of the arch. Further extension of the compass will scribe the extrados of the arch from point 3.

Alignment of the units of the arch is from point C, where the center line bisects the spring line. Should the alignment be from radial point 7, the abutments on which the springer at the skewback units sit should be in alignment with the radial point 3. There is less work shaping units when the alignment is from the radial point 7, which is used to scribe any arch. Also, the units will be more uniform in size.

The skewback is found by striking a line from the radial point 7 through points 2 and 3 to the extrados. All units of the arch are aligned in relation to radial points (Fig. 8-21).

158

Fig. 8-19. A semi-Gothic arch.

159

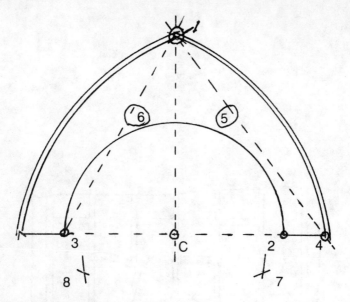

Fig. 8-20. Alignment for the semi-Gothic arch.

Elliptical Arch

An *elliptical* arch is a very strong arch. It springs from a horizontal seat at the spring line.

There are several methods of laying out the elliptical arch: by string, template trammel, compass, or other means. When using the string method, take one-half the distance of the span of the opening in the compass, place the point of the compass as point 1, (this arch rise should be less than one-half the span line), and cut arcs on the spring line at points 4 and 5. Drive nails at points 1, 4, and 5 on the layout deck or panel. Tie the end of a string to the nail at point 4, pass it around the nail at point 1, and tie to the other point 5. Replace the nail at point 1 with a pencil, and scribe the elliptical arch from point 2 to point 3. By lengthening the string, the extrados may likewise be scribed.

In order to arrive at the units' size, shape, and placement in this arch, determine the units' coverage at the extrados. Caliper the units' size with the compass, allow for the mortar joint, and mark off the extrados line. The units need sizing to fit properly. This unit spacing is aligned with various radial points on the center vertical rise line extended below the spring line. Strike off the alignment

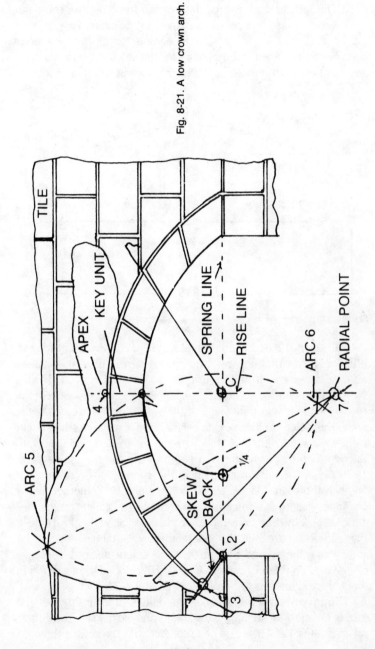

Fig. 8-21. A low crown arch.

161

lines and chip, cut, or rub down the units to fit each space between extrados and intrados lines. Always let the key center the arch.

The face breadth, which determines the length of the units, varies regarding the necessity for strength and desired appearance. Space devisions must be such in number to allow the longest member to be placed at the intrados for the key and the first units at the base of an arch that springs at the skewbacks (Fig. 8-22).

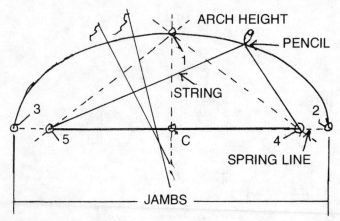

Fig. 8-22. The string method of making an elliptical arch.

Trammel Rod. The trammel rod is needed when making an elliptical arch with a square or trammel square. This rod is marked off and holes are drilled according to the arch desired. Having located the center C of the arch's span, mark and drill a hole in the rod, which is longer than one-half the span plus the breadth of the arch. This hole should have a dowel inserted in it which sticks out at least enough to come in contact with the square or fit securely into the depth of the perpendicular rise groove in the trammel square. Drill a hole the size of a pencil in the rod using the distance of one-half the span. Place a pencil in this hole. Drill another hole a distance equal to the breadth of the arch desired and insert another pencil. Mark the first hole 1, the second hole 2, and the third hole 3. Place the rod perpendicular in alignment with the rise line of the arch. Place the second hole at the arch height desired. Measure from 2 to C on the rod, mark at C, and drill a hole. Drive a dowel in this hole and mark it 4. See Figs. 8-23 and 8-24.

Square Method. The square method is an accurate method if care is taken when scribing the intrados and extrados of the elliptical arch. Pins 4 and 5 should be in contact with the square through-

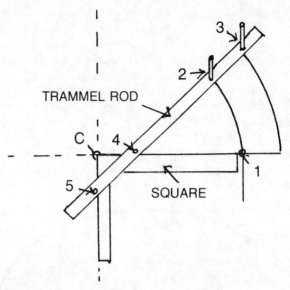

Fig. 8-23. The square method of making an elliptical arch.

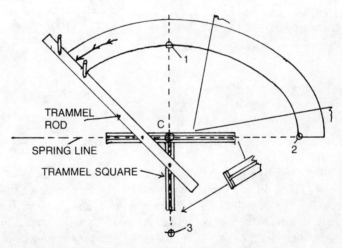

Fig. 8-24. The trammel method of making an elliptical arch.

out the forming of this arch. The square should be held firmly in place (Fig. 8-23).

Trammel Method. The trammel square method is a more secure and accurate method to use when making an elliptical arch (Fig. 8-24).

Compass Method. There are two compass methods I have used. The elliptical arches made with these methods are not true elliptical arches. They should be plastered.

In the first method, the span is marked 2 and 3, respectively. The center of the span, C, is crossed with the extended rise line above and below the spring line. Divide the spring line into four equal parts to find 4 and 5. With the compass opened the distance from 2 to 4, place the compass point at 2 and 4 to find the arc 6 above the spring line. Draw a line from this arch through 4 and cut the rise line 8. The radial points of this arch are 4, 5, and 8. From these radial points, the intrados and extrados can be scribed with the compass (Fig. 8-25).

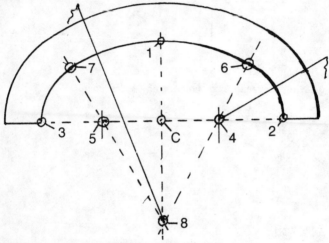

Fig. 8-25. The compass method of making an elliptical arch.

The second compass method used when scribing an arch is as follows. Place the spring line and the rise line and mark on the spring line the span distance at 2 and 3, which is an equal distance from C. Designate the arch height 1 as desired (this height should be less than one-half the distance from 2 to 3). With the compass point at 2, having a distance from 2 to C, cut off a straight line from 1 to 2 and mark it 6. With the compass point at 2 and 6, having a distance on the compass greater than one-half the distance from 2 to 6, the arcs 8 and 9 are cut and found. Draw a line from 8 through 9, and cut the rise line below the spring line to find the radial point 14. To check, take the balance of the line distance from 1 to 2, that is, the distance from 1 to 6 in the compass. With the point at 2, cut the spring line to find 4. The distance from 4 to C on the compass with the point at 2 and 6 arcs below the spring line and cuts an arc on the line from 8 to 14. The arch haunch parts of this half of the elliptical arch can be scribed from radial points at 12 and 14. The other half of the arch can be made the same way (Fig. 8-26).

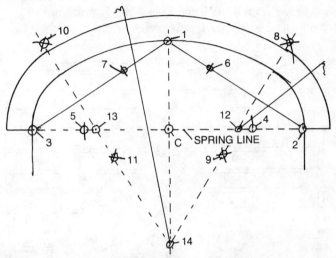

Fig. 8-26. Another compass method of making an elliptical arch.

Ogee Arch

The *ogee* arch is not a strong arch, due to its S-haunch. It does have a unique appearance. There is more compass work in designing this arch than in any other arch.

Having struck the spring line and the rise line, mark the cross center of the arch span C. Designate the width of the arch on the spring line as 4 and 5, respectively. Determine the breadth faces of the arch and mark them 2 and 3. To find the apex 1 of the arch, place the compass at C and extend it to 2 plus one more breadth of the arch face. With this distance in the compass, cut the rise line above the spring line and designate this apex point as 1. Strike a line from 1 to 2 and from 1 to 3. Cut these lines in half and designate them as 6 and 7, respectively.

To find the radial points, place the compass point at 6 with the distance in the compass from 2 to 6. Strike an arc from 6 and 2 and designate the arc as 9. Do the same from 3 and 7 and designate the arc 8. With the compass, having the same distance, strike in a like manner, using 1 and 7 and 1 and 6. Designate the points 10 and 11, respectively. Strike a line from 8 to 11 and from 9 to 10. These lines cut through 6 and 7. From arcs 8 and 9, the intrados and extrados can be scribed from 3 and 5 to 7 in alignment from the arcs 8 and 10. The same can be done from 4 and 2 to 6 at the straight line which extends from 9 to 10. The upper part of the arch can be scribed in a like manner from the radial points at arcs 10 and 11, using the distance in the compass from 6 to 10 and from 7 to 11, which are the same distances as 1 to 10 and 1 to 11, respectively. See Fig. 8-27.

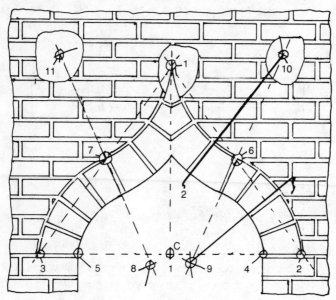

Fig. 8-27. An ogee arch.

Egg

An *egg* is not too difficult to install if a movable buck form or centering is used. The upper part of the egg from the spring line is a semicircular arch scribed from the center of the spring line at point C.

When laying out the egg on the layout deck, the extended spring and rise lines are struck at right angles to each other. Having decided on the height of the arch above point C, place the compass point at point C. Scribe the intrados of the upper part using the distance from 1 to C, and also cut the spring line to find 2 and 3. Take one-half of the span opening on the spring line in the compass. Extend the compass to scribe the extrados. By taking the span

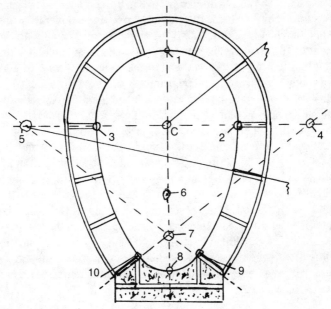

Fig. 8-28. An egg.

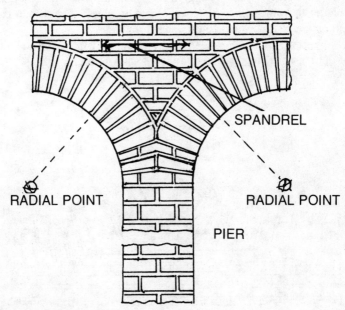

SPANDREL

RADIAL POINT

RADIAL POINT

PIER

Fig. 8-29. A springer arch.

distance along the spring line from 2 to 3 in the compass, and placing the compass point at 1, cut the extended rise line below the spring line and find 6 (Fig. 8-28). Using the distance from 2 to 3 with the point at C, the point at 8 is found by cutting the rise line below the spring line. Point 7 marks one-fourth the rise line from 8 to C.

Points 4 and 5 are found by using the distance from point 1 to point 7 with the compass point at point 1 and point 7, respectively. With the compass point at point C and extended to point 1, the intrados of the upper part egg can be scribed. With the compass at point 4 and 5, the intrados of the egg from 3 and 2 to 9 and 10, respectively, can be scribed. The points found by lines 4 and 5 through 7 are the arc's intersections with the lines from 4 and 5 through 7. With the compass point at 7 and opened to point 8, the intrados at the bottom can be scribed from point 9 to 10. In like manner the extrados of the egg can be run by opening the compass the distance of the breadth of the egg's face. The units are in alignment with their respective radial points. Because the egg is generally used as a sewer, all mortar for the units must be sement mortar.

Springer Arch

The *springer* arch springs from a column to a wall, another column, or from a wall to a wall or column to column. The spandrel is a filler between or next to the haunch of an arch.

A spring arch has a tendency to push out the jambs. The method of uniting springing arches to a pier is shown in Fig. 8-29.

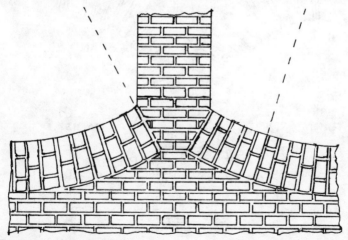

Fig. 8-30. An invert arch.

168

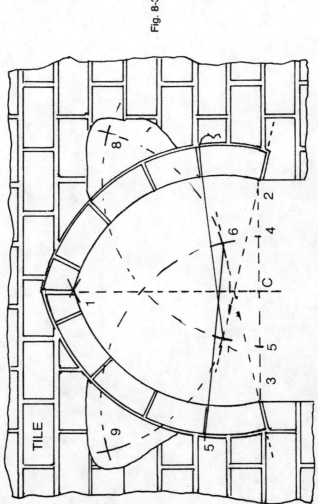

Fig. 8-31. A horseshoe arch.

Invert Arch

An *invert* arch receives stress pressure. The method of placing the units in this arch is shown in Fig. 8-30.

Horseshoe Arch

A *horseshoe* arch continues below the spring line. It is two segmental arches leaned together at the apex. It is laid out as is the segmental arch (Fig. 8-31). The horseshoe arch, since it is shaped like a horseshoe and is curved in at the bottom, has a tendency to push at the jamb at the impost.

Chapter 9

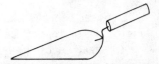

Steps, Patios, and Walkways

Masonry steps should be reinforced with re-bar whether construction units or concrete are used. Steps can be built with a combination of units providing the mortar used is waterproof and free from voids in its construction. The joints can be tooled or brushed.

STEPS

Steps should have a slope for drainage of ⅛″ to 4′ run. The re-bar should not be over 12″ apart and should be placed crossways as well as lengthways, and up and down. The bars should be at least ⅜″ (Fig. 9-1).

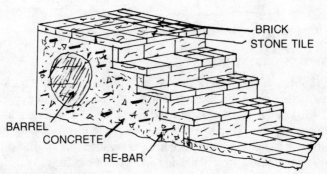

BRICK
STONE TILE

BARREL
CONCRETE
RE-BAR

Fig. 9-1. Masonry steps.

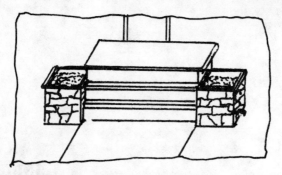

Fig. 9-2. Steps can be made with planters on both sides.

Steps can be made with planters on one side or on both sides. Rails or sides of steps can be built along with the steps. Care should always be taken not to leave any cracks or places where moisture can enter and ruin the masonwork (Fig. 9-2). Trim can be carried up above the steps in various ways (Fig. 9-3).

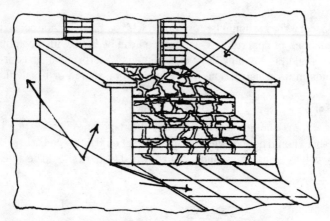

Fig. 9-3. There are many ways of carrying trim above the steps.

PATIOS AND WALKWAYS

Patios and walkways can be built with any of the various construction units. For a cheap and usable patio or walkway, the units are laid on a bed of gravel and sand mix. The joints are filled with sand or dirt. Grass can be planted in the dirt joints. A good permanent patio or walkway requires excavation, drainage, a good base, and solid construction using waterproof mortar or concrete (Fig. 9-4).

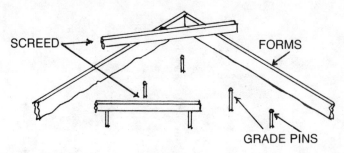

Fig. 9-4. Take time when excavating to grade.

A patio or walkway should have a slope of ¼" to 4' for drainage. Drainage should be provided for the underbed. Water can freeze and heave up the units. Re-bar or a good 14 gauge 6" mesh welded galvanized wire should be used. All joints between the units should be filled with a good cement mix and be free from voids. Should water in voids freeze, the joints will come apart. All patio or walkway units should be level with each other. Use a straightedge or screed for leveling purposes (Fig. 9-5).

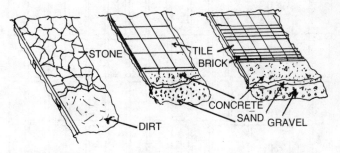

Fig. 9-5. Walkway and patio detail.

There are many patterns for laying a patio or a walkway with bricks. The best bricks to use are hard bricks. These bricks do not readily absorb water and wear well. Bricks can be laid either flat or on edge. They last longer and stay in place better when laid on edge than when laid flat with their wide width up.

Bricks can be laid in dirt with dirt joints. The dirt on which bricks are laid should be packed. When bricks are laid on concrete, good cement mortar should be used as an underbed and in the joints (Fig. 9-6). Another brick walkway, in a different bond, is shown in Fig. 9-7.

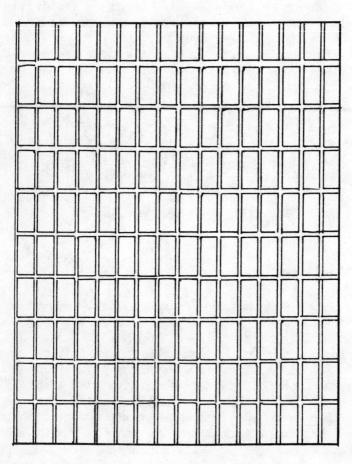

Fig. 9-6. A walkway with bricks laid flat.

174

Fig. 9-7. Another brick walkway.

175

Fig. 9-8. A herringbone brick bond patio.

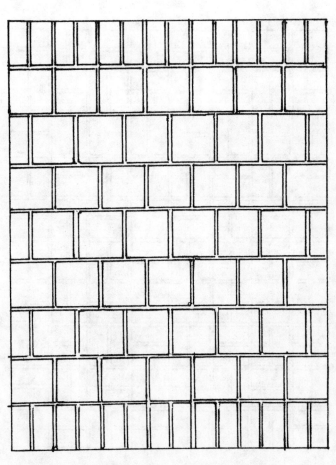

Fig. 9-9. A brick and tile walkway.

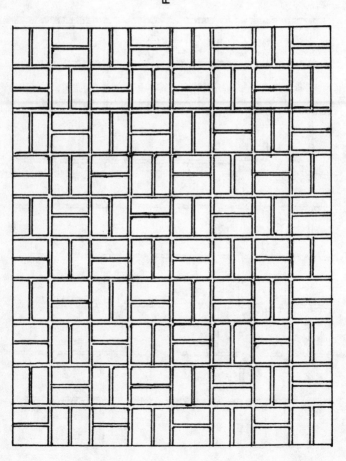

Fig. 9-10. Another brick bond patio.

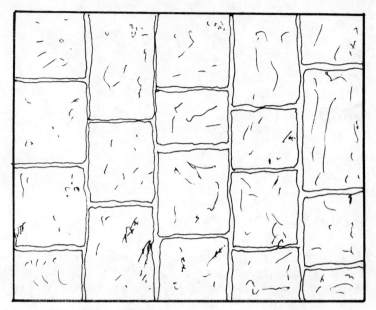

Fig. 9-11. A walkway laid with stone.

There are many sidewalks laid using the *herringbone* bond. Take the time to lay out the brick without mortar first. See Fig. 9-8.

Combinations of the various masonry units in the construction of patios and walkways make pretty masonwork. Figure 9-9 shows a combination using brick and outdoor tile. Bricks laid in blocklike sections make a pleasant bond in a patio (Fig. 9-10).

Stone patios and walkways are beautiful and long lasting. Stone can have a pattern or be set in the patio randomly. Figure 9-11 shows a pattern of stone in a walkway.

Chapter 10

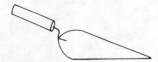

Fireplaces

Faulty construction rather than shape, size, style, or units used is the prime reason that a *fireplace* works improperly. A fireplace that smokes from the wrong end is useless, dirty, and costly when smoke damage must be repaired.

A fireplace should have a solid *footing*. It should not settle, crack, or come apart. The material used for a footing should withstand the weathering elements to which it may be subjected.

The fireplace should be well reinforced with re-bar, dur-o-wall, metal ties, etc. Perpendicular re-bar at corners should extend into the footing. This keeps the fireplace from cracking and falling apart. A cracked fireplace may smoke where it should not. All trimmer arches or overhangs of a fireplace should be reinforced, whether of concrete or construction units.

Allowance for expansion is necessary when constructing a fireplace, whether or not fireplace units or dampers are used. A 1″ space allowance between the unit or firebrick and the fireplace proper walls is a must. A unit should have insulation separating it from the masonwork (Fig. 10-1).

Firebox linings accommodate any expansion of firebox units. All firebox firebrick units should be laid in fireclay mortar. This mortar is of thin grout consistency, and the units are dipped in it and set.

FIREPLACE PARTS

The various parts or divisions of a fireplace are shown in Fig. 10-2. Fireplace footings should be reinforced and extend beyond the fireplace proper at least 8″ (Fig. 10-3).

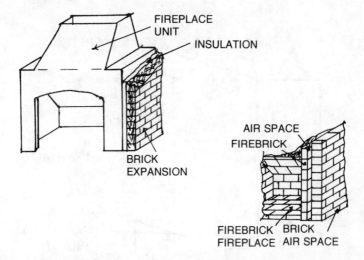

FIREPLACE UNIT

INSULATION

AIR SPACE
FIREBRICK

BRICK EXPANSION

FIREBRICK BRICK
FIREPLACE AIR SPACE

Fig. 10-1. Insulation should separate the unit from the masonwork.

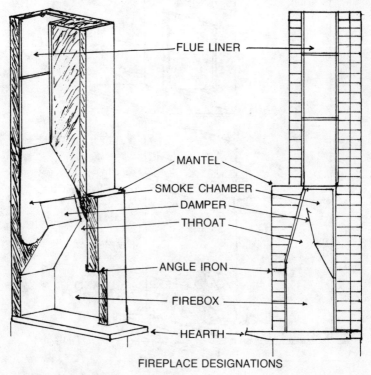

FLUE LINER

MANTEL

SMOKE CHAMBER

DAMPER

THROAT

ANGLE IRON

FIREBOX

HEARTH

FIREPLACE DESIGNATIONS

Fig. 10-2. Fireplace parts.

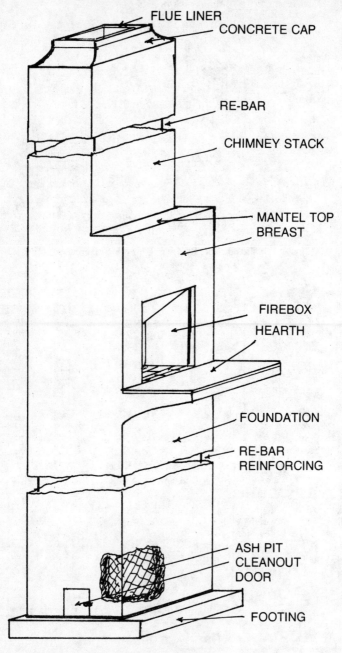

FLUE LINER
CONCRETE CAP
RE-BAR
CHIMNEY STACK
MANTEL TOP
BREAST
FIREBOX
HEARTH
FOUNDATION
RE-BAR
REINFORCING
ASH PIT
CLEANOUT
DOOR
FOOTING

Fig. 10-3. Another fireplace parts diagram.

182

FIREPLACE PROPORTIONS

There are various proportions to consider when laying a fireplace. The area of the opening in the front face of a fireplace breast is the base unit. The height of this opening is determined by the height of the chimney stack. A 24″ opening is suitable for a 20′ chimney stack when the opening is 24″ wide. Any fireplace opening can be less in height than is stated, but it should never be more. When the chimney stack is 24′ high, the opening may be raised 1″. If the chimney is 16′ high, the opening should be lowered 1″. Width can equal height. The chimney stack of a fireplace should extend at least 3′ above a 10′ horizontal point of anything that is above it. The part of a fireplace that extends outside should have at least 8″ thick walls.

The area of a flue liner should be at least 12 percent of the base unit area. This may vary with the height of the chimney stack. When the chimney stack is 26′ high from the hearth, the flue liner area could be reduced to 11 percent of the unit area. If the chimney is 14′ high, the flue liner area could be 13 percent of the unit area.

Firebox depth should not be less than two-thirds the width of the unit opening width (4 in Fig. 10-4). The back width of a firebox

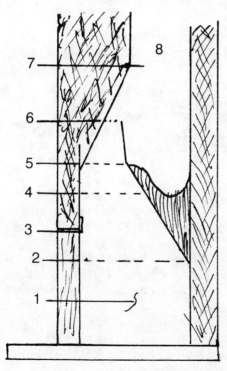

Fig. 10-4. Fireplace divisions.

should be one-third less than the unit opening width. The slope of the firebox sides and back to the damper should not be more than 40 degrees from the vertical.

Damper opening should be equal to two-thirds the area of the liner. The damper plate should be at least long enough to extend over and not hinder smoke entering the smoke chamber (4 in Fig. 10-4).

The distance from 3 to 5 in Fig. 10-4 should not be less than one-third the height of the opening. The damper plate (6 in Fig. 10-4) should not open closer than 1" to the vertical of the inside face of the liner. The bottom of the smoke shelf or smoke table (4 in Fig. 10-4) should be below the level of the bottom of the damper.

The degree of pitch from 5 to 7 in Fig. 10-4 should not be more than 30 degrees from perpendicular. The angle iron (3 in Fig. 10-4) should extend at least 4" into the masonry at its ends and should have insulation at the ends for expansion. The hearth in the firebox should be of firebrick. When a fireplace unit is not used, the fireplace from and including the firebox should have firebrick laid on edge to 7 in Fig. 10-4.

FIREPLACE LAYOUTS

Decide the best place for a fireplace in a room. The fireplace layout can then be marked out. One common layout is shown in Fig. 10-5.

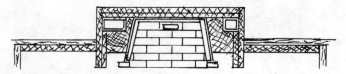

Fig. 10-5. Fireplace layout.

Different fireplace layouts are necessary when building fireplaces in certain locations. A corner fireplace can be set in or out as well as a wall fireplace (Fig. 10-6). Figure 10-7 depicts a fireplace opening on the side.

A corner layout can have the back of the fireplace extend outside the house, or it can be inside the house (Fig. 10-8). A mantel for a fireplace is optional (Fig. 10-9). Fireplaces can be built that open on both sides (Fig. 10-10).

A fireplace that opens on both sides and end is shown in Fig. 10-11. A fireplace that opens on all sides should be reinforced well. The corners of the fireplace are the stabilizers (Fig. 10-12).

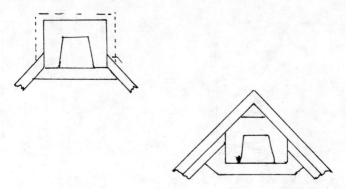

Fig. 10-6. Corner fireplace layouts.

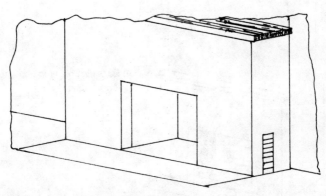

Fig. 10-7. A fireplace opening on the side.

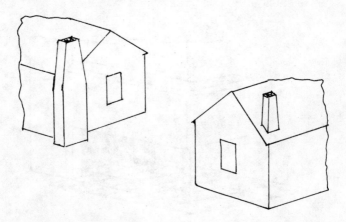

Fig. 10-8. More corner layouts.

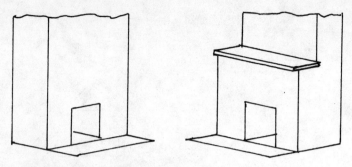

Fig. 10-9. A mantel for a fireplace is optional.

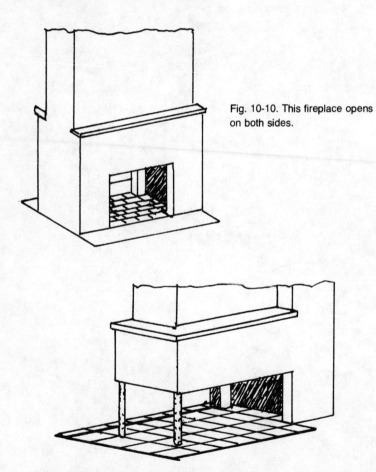

Fig. 10-10. This fireplace opens on both sides.

Fig. 10-11. The fireplace opens on both sides and the end.

186

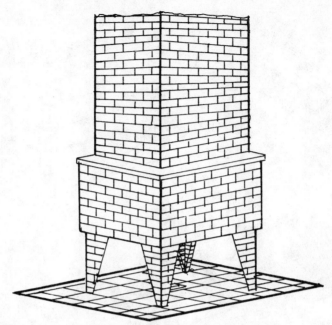

Fig. 10-12. The corners of the fireplace are the stabilizers.

The hearth should be either level with the floor or raised. A trimmer arch is used to hold up the extended hearth when a concrete slab is not used (Fig. 10-13).

A fireplace should have access to oxygen. Vents are installed that bring in fresh outside air (Fig. 10-13).

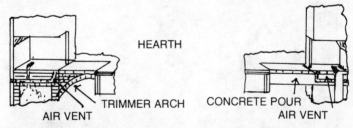

HEARTH

TRIMMER ARCH

AIR VENT

CONCRETE POUR

AIR VENT

Fig. 10-13. Note the vents.

CHIMNEY STACKS, DOUBLE
FIREPLACES, AND BACK TO BACK FIREPLACES

A *chimney stack* can have various tops (Fig. 10-14). The straight convex and concave mortar on the top can either retard or enhance the drawing of a chimney stack (Fig. 10-15).

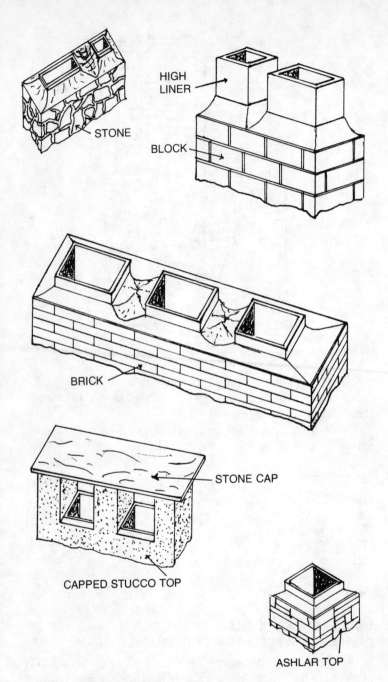

STONE

HIGH LINER

BLOCK

BRICK

STONE CAP

CAPPED STUCCO TOP

ASHLAR TOP

Fig. 10-14. Chimney stack tops.

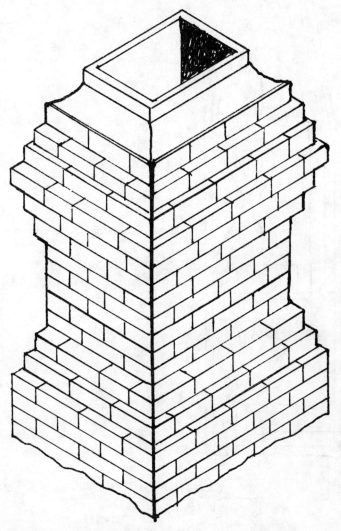

Fig. 10-15. A basket or corbeled stack.

Fireplaces can be single, double, or stacked (Fig. 10-16). Openings can be on one side, both sides, the end, or on one side and the end.

Back to back fireplaces can use the same chimney stack (Fig. 10-17). Two or more fireplaces built together should have separate chimney liners.

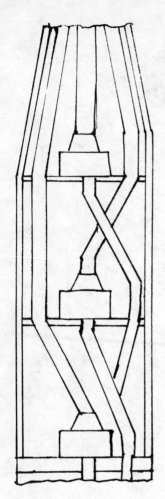

Fig. 10-16. Fireplaces can be stacked.

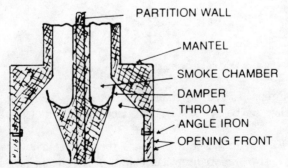

PARTITION WALL

MANTEL

SMOKE CHAMBER

DAMPER

THROAT

ANGLE IRON

OPENING FRONT

Fig. 10-17. Back to back fireplaces can use the same chimney stack.

FLASHING

Flashing is required at the roof line. The fireplace chimney stack should have from 1″ to 2″ clearance from the roof. This stops the house vibrations from carrying to the fireplace or chimney stack.

FIREPLACE SUPPLIES

A fireplace unit is shown in Fig. 10-18. Fireplace supplies include an ash dump, ash cleanout door, damper, angle iron, and reinforcing (such as dur-o-wall, brick tires, and re-bar).

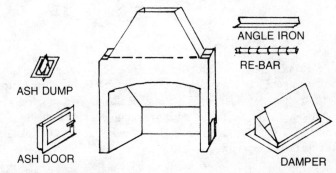

ASH DUMP

ASH DOOR

ANGLE IRON

RE-BAR

DAMPER

Fig. 10-18. A fireplace unit.

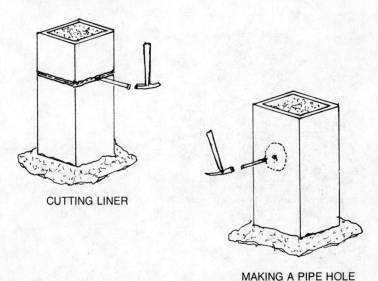

CUTTING LINER

MAKING A PIPE HOLE

Fig. 10-19. Here are the ways to cut a flue liner and make a hole for a pipe.

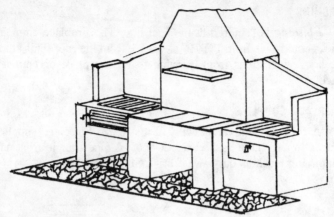

Fig. 10-20. An outdoor barbecue.

A fireplace can be built with any of the units of construction. Firebrick should be used next to the fire and in the throat, fire hearth, and smoke chamber. Firebrick can be dipped in a fireclay grout and set at once. The other units require a mortar made of 1 part cement, ¼ part lime, 3 parts good graded sand, and drinkable water mixed to a plastic consistency.

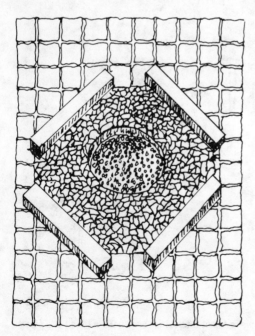

Fig. 10-21. A fire pit.

Flue liners are required in zoned construction areas. The easiest way to cut tile is with a saw. The method of making holes and cutting flue liners is shown in Fig. 10-19.

BARBECUES

Outdoor *barbecues* are made in different sizes and shapes. Many outdoor barbecues are built with scrap or broken masonry units put together with or without mortar. The beautifully built ones are made with care (Fig. 10-20).

A fire pit can do if a barbecue is not at hand. A removable grill can be placed over the fire pit. Almost anything can be used for seats (Fig. 10-21).

Chapter 11

Walls and Planters

Garden walls can be built with different masonry units and in various sizes and designs. A good footing is necessary. Some walls that are straight and without corners should have their footings extend well into the ground. Otherwise, they eventually fall over. Patio walls can be high enough to hold a *portico* roof. Yard walls need not be high (Fig. 11-1).

WALLS

The concrete posts extending into the ground keep the wall in place (Fig. 11-2). A stone garden wall looks strong (Fig. 11-3). All garden walls need caps. They are built to run off water and keep the top of the wall from absorbing moisture (Fig. 11-4).

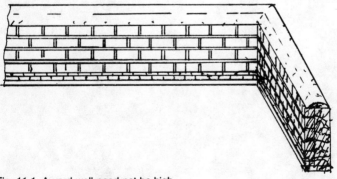

Fig. 11-1. A yard wall need not be high.

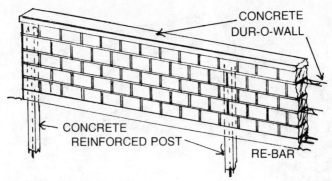

Fig. 11-2. Note the posts.

Walls can be made with an artistic touch. Figures 11-5 through 11-8 show some designs.

PLANTERS

Planters can be built with all the masonry units. They can be small, large, low, or high. Planters must last without cracking or coming apart. To keep the frozen dirt in a planter from swelling to the point of bulging out the walls, the inside should be cupped. This allows the wet dirt in the planter to rise up instead of bulge out (Fig. 11-9).

Another way is to have the bottom part filled with rock for drainage. A pipe can be used or an opening left for drainage (Fig. 11-10).

Low planters do not crack as readily as high ones (Fig. 11-11). Planter footings should be reinforced (Fig. 11-12).

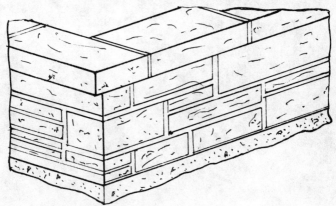

Fig. 11-3. A solid stone garden wall.

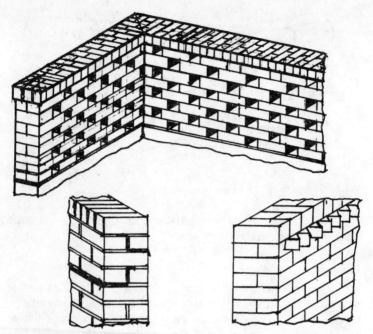

Fig. 11-4. Stone garden walls need caps.

Fig. 11-5. A wall with an attractive gate.

Fig. 11-6. An attractive wall.

Fig. 11-7. Another eye-catching wall design.

Fig. 11-8. A carefully constructed wall.

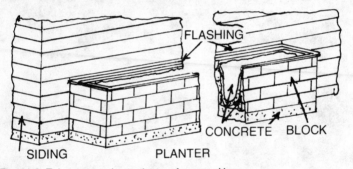

Fig. 11-9. Take steps to keep planters from cracking.

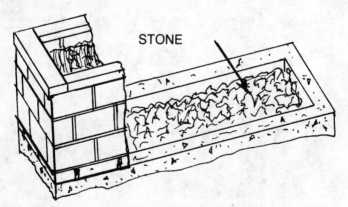

Fig. 11-10. The bottom part of the planter is filled with stone.

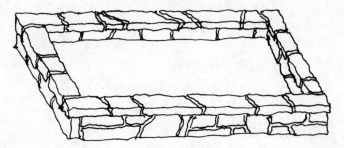

Fig. 11-11. Low planters are more resistant to cracking.

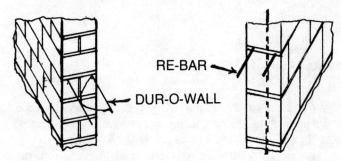

RE-BAR

DUR-O-WALL

Fig. 11-12. Plaster footings should be reinforced.

Chapter 12

Contracting

I have been a contractor for more than 30 years. I joined the bricklayers and stonemasons' union and have been a member for more than 30 years, also. I should know the construction business.

Anyone can contract. If you can estimate the requirements and materials of a job, you will be a successful contractor. If you do not, you will not be successful.

To be a masonry contractor, you should have a knowledge of mathematics. You should understand the decimal system. The metric system is now being used in the United States, also. You will not stay in the business of contracting masonwork very long unless you can figure. The most important thing in contracting is the ability to estimate correctly, for estimating a job is nearly all figuring.

To contract masonwork, you should be familiar with the terms used in construction. You should know the names and uses of masonwork materials. You must be able to operate the tools used in masonwork.

You have to have good credit or capital on which to operate. Therefore, your best assets as a contractor are honesty and the ability to keep every promise. Dependability and promptness are necessary for every contractor to consider. These are things, along with integrity and good judgment, that will help you in your contracting business.

Many beginning masonwork contractors lack business experience and the ability to estimate the cost of a masonwork job. Many of them have been bricklayers, foremen, superintendents, or have worked in building construction. Many of the masonry contractors without business training go broke. They should have had business experience or capital to tide them over until they gained through contracting experience the knowledge a contractor should have.

Generally, contractors without business experience take jobs at such a low price that they go broke unless they do much of the work themselves. They barely make a living and would be far better off working for wages.

It is the low bidders, in most cases, that are forced to do inferior work to stay in business. These are the ones who ruin not only the contracting business of competent builders, but ruin themselves. When contractors do inferior work to stay in business long enough to learn the business of estimating, they likely go out of the masonwork business for lack of jobs.

BUSINESS ACUMEN

Stonework contracting is a business. Business in any field should be carried on in an efficient manner. You should keep records of all masonwork jobs. Quantities and costs of materials and where they can be obtained should be recorded. Cash discounts should be noted for future jobs. Estimates of all jobs should be kept. The names and addresses of workers should be listed for future hiring. All transactions and blueprints, along with the specifications of completed jobs, should be kept for reference. Receipts of all expenditures should be kept.

There are so many things to consider before you actually enter the business of masonwork contracting. You have to keep books and records if you are going to be a small contractor. If you want to be a big contractor, you should have an office and everything that goes with it. You will need office help, especially if you cannot tend to the office yourself. Your overhead expense will become a stable, fixed expense and may, percentagewise, enter into all job estimates, or it will have to come out of your yearly profit.

EXPENSES

Office rent and upkeep, heat, lights, office equipment and supplies, fixtures, telephone, wages for office help, fire and liability insurance, social security, taxes, and unemployment compensation are necessary expenses.

Magazines, books, club dues, association dues, advertising, traveling expenses, fares, automobile and truck expenses, and possible legal expenses are generally forced on contractors.

Yearly expenses include contractor's salary, estimator's license and salary (unless the contractor does his own estimating), storage rents, general tool equipment upkeep and repair (which cannot be charged off to a job), interest on contracting investment, etc. There is also the help cost necessary to carry on the contracting business which cannot be charged off on any particular job. The cost of a contractor's license is another consideration.

You must decide whether to hire union or nonunion employees. If a union employer, you should understand the union regulations with which you are required to comply. You should know and understand your subcontractors. Should you sublet parts of your job? Will you have to carry a subcontractor's bond as well as yours?

ESTIMATING

Correct estimates are a must when you are contracting masonwork. If you cannot estimate, it is to your advantage to have office employees that can carry on this particular end of contracting. You should learn to check their work carefully.

To estimate properly, you should understand the specifications, know how to read and check the measurements on the blueprints or plans, and be able to list kinds and quantities of material needed. You should know where the material is needed in the job, the time it takes to put it in the job, and the cost of putting it there. A properly prepared estimate, the purchasing of materials, the careful way you sublet (if you sublet), and the way you manage the work on the job are all important things to take care of if you want to stay in business and make a profit. You then have figured everything pertaining to the job in detail. Your bid will be correct, and you will know exactly how you will stand on any job you may get, barring, of course, weather and strikes.

Labor costs should be listed along with material costs. The reason is that you need a reference to consider when bidding on future jobs. You will know just how well your estimates hold up after a job is completed. You can then take that into consideration when bidding on jobs. The cost knowledge of the time it takes to perform a certain part of a job, taking into consideration the materials and conditions, is generally learned by cost records of previous masonwork jobs performed.

Job conditions, temperature, the lay of the site, weather, room to place materials, working room, and access to the job should be taken into consideration when estimating. Always make an allowance for inconveniences which may increase the cost of a job.

Job costs should contain only a percentage of overhead expenses based on the total yearly overhead expense you may have during the year. The following costs should be in your estimate when figuring a job: building permit; survey of site cost; temporary roads; repair of any destruction which may happen during construction; protection of work (finished and unfinished); removing and building sheds, shelters, etc.; removing and replacing masonwork; cleaning and removing rubbish, bucks and platforms; all temporary work such as decks; all carpenter expenses; trucking; drainage; runways; scaffolding; bonding; social security and unemployment taxes; sales taxes; fees; fuel; electricity; breakage; timekeeper; tool man, watchmen; water man; legal expenses; insurance; labor; superintendent; foreman; layout man, etc. Don't forget contingencies.

One thing you should consider very carefully is profit. That is the main reason for any contractor to be in the business. The total cost plus profit should include contingencies. If the amount for contingencies is too large, you may not get the job. If your estimate is too low and your bid is low, then if you get the job, you will certainly be short on your contingencies. Then, too, high bidders will not get a job unless they have built up a reputation of honesty, dependability, service, and good work.

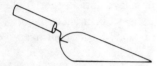

Glossary

abutment—Part of a structure that directly receives thrust or pressure of an arch, vault, beam, or strut.

acute angle—An angle less than 90 degrees.

additive—Material added to an agglomerated mass for a specific purpose.

adhesive strength—The strength mortar has for holding two stones together.

agglomerated mass—Matrix or binder and the aggregate of a mix such as plaster, mortar, and concrete.

aggregate—Hard, inert material in various size fragments mixed with cementing material to form concrete, mortar, or plaster. Particles of an agglomerated mass other than the binder and additives.

alignment—Construction units laid true to the line.

amalgamation—Mixing of various materials together in a unified mix such as lime, cement, sand, color, and water.

American bond—Masonry bond in which headers recur every fifth or sixth course, stretcher courses laid to break joints. Also called common bond.

anchor—Metal rod or strap tie that gives stability to one part of a structure by making it fast to another part.

anchor grooves—Grooves cut in construction units to hold anchors.

angle bar—See anchor.

angle iron—Iron or steel cleat or brace used to hold together two parts whose faces are at an angle. It is used to hold lintel units.

anta—Pier produced by thickening a wall at its termination. A pilaster attached to a wall.

apprentice—Person bound by a contract to learn a trade.

arcade—Series of arches supported by columns or piers.

arch—Curved structural member spanning an opening and resolving vertial load pressure into horizontal or diagonal thrust.

arch buttress—See flying buttress.

area—The surface within any set of lines usually given in square units of measure.

arris—Sharp edge or angle formed by the meeting of two surfaces, whether plane or curved.

artificial—That which is not natural; man-made or synthetic.

ashlar—Squared and dressed stone; also, masonry of such stone.

ashlar line—Exterior line of an exterior wall above any projecting base.

axed work—Incisioned successive rows ⅛ to ¼" apart across a stone.

axis—In masonry arch work, the center or point from which a circle, arc, or intrados (and extrados) is formed or drawn on a template form.

backing—That part of the wall behind its face.

backing up—Laying the backing.

backset—A construction unit or course set back from the face of masonry work.

back side—Opposite from where the mason stands.

back work—Laying up the back side before the front side is laid up; generally 2' high, so it can be parged or plastered.

balanced—An even distribution of the load.

basement—Lowest fundamental part of a structure that is above a footing; wholly or partly below ground level.

bat—Brick with one part whole and the other broken off. The structurally usable part of the brick.

batter—Deviation from the vertical.

batter boards—Single or double boards placed outside the corner of an intended structure from which true lines are run.

batter stick—Tapered or pitched stick used with a level to build a battered wall. A tapered board attached to a spirit level.

beam—Main horizontal load-bearing member supported by walls, columns, etc.

bearing stone—Stone supporting a load other than the units of which it is a part in the masonwork.

bearing wall—A wall or partition which supports weight.

bed—The top surface on which mortar is spread.

bed joint—Horizontal joint between construction units.

bed surface—Unfinished surface that carries the finish.

belt course—Horizontal course or band around pillars or engaged columns and brickwork.

bench mark—Elevation mark or point of reference on a fixed object.

binders—Material such as cement or lime that produces cohesion of loose material such as aggregate.

blind bond—Masonry bond in which headers extend only halfway through the tier of stretcher face brick; bonding in a wall that does not show on either side.

block—Construction unit dressed to uniform size.

block and bond—Combination method of laying a block bond on one side and a different unit and a bond on the other.

blocking course—Stone laid on top of a cornice to give weight that binds the cornice in place.

boasted work—Stone dress made with a boaster (chisel with a wide cutting edge). Boasted marks are not kept in continuous rows across the stone.

bond—Systematic lapping of construction units in the face of masonry for strength and appearance.

bonding course—Course in which construction units are laid transversely part way or entirely through a wall from the face.

boning rods—Rods to sight across to keep construction units level when laying them on a horizontal surface.

breadth—The face height of a stone in stonework.

breaking joint—Method of laying so no two vertical joints will be immediately over one another.

breast of a fireplace—The front of a fireplace.

breast wall—Wall built to sustain the face of a natural bank of earth.

brick—Construction unit.

brick-and-brick—Brick laid end to end with very little mortar, if any.

brick veneer—Brick facing on wall built of other material.

brickwork—Laying brick with mortar. The finished product.

bridge wall—Wall separating firebox from combustion chamber in a boiler setting.

broken range—Random laying; no true bond kept or desired.

brooming—Finishing concrete with a broom.

buck—Rough, well-braced frame placed in a wall during construction.

buckstay—Either of two connected girders, one on each side of a masonry structure to take the thrust of an arch.

bulging—Spreading or widening of a wall or pour generally caused by load pressure.

bull header—Brick laid on edge with the end showing, as in a rowlock course.

bull stretcher—Brick laid on edge with its broad side showing.

bull's-eye—Circular opening in a wall.

bush hammer—Hammer which is used for dressing stone and concrete.

buttered joint—Thin masonry joint made by mortaring laid brick; generally, mortar does not fill the head joint by using this method of laying and by mortaring brick on one end.

buttress—Projecting masonry structure that resists lateral pressure at a particular point in one direction.

calcining—Heating stone for manufacturing cement or lime.

camber—Slight arch or curvature.

carving—Dressing a stone or stonework to a carved finish.

casting—A molded or cast construction unit.

catface—A low spot on a plastered or stucco wall.

caulking—Filling shrinkage cracks around window frames and expansion joints to make them weathertight.

cavity wall—Masonry wall built in two thicknesses separated by an air space; double-wall construction.

cell—Air space in a construction unit; a cellular cavity.

cement—A powder of alumina, silica, lime, iron oxide, and magnesia burned together in a kiln and finely pulverized; when mixed with water, it forms a plastic mass that hardens by chemical combination.

center—Temporary prop or support similar to an underpin.

centering—Framework used temporarily on which a mason may apply a compass, template, or trammel rod when laying out arches. Timber falsework is part of the masonry arch during construction.

chain course—Bond course of headers continuously fastened together by cramps.

chase—Groove or channel left in masonwork for a pipe or conduit.

chimney—Hollow masonry stack that carries off smoke.

chimney breast—Projection of a chimney or fireplace from a wall into a room or projecting on the exterior of masonry.

chimney cap—Uppermost course or unit of a chimney stack. It is used to improve the draft by presenting an exit aperture to leeward.

chimney lining—Parget or flue liner.

clastic—A conglomerate stone could be classified as a clastic stone. It is made up of other particles of stone, as is granite.

cleavage plane—Natural division of stone in that two parts of the same stone separate along the cleavage plane. It is in reality a seam in stone.

clip bond—Masonry wall bond formed by clipping off the inner corners of the masonry units; used to unite diagonal bond with a stretcher face.

closer—The last construction unit laid in a course to close the course and maintain the bond.

coats—Layers of plaster work (as in one-coat work, two-coat work, three-coat work, and finish coat).

cobblestone—Naturally rounded stone, larger than pebbles and smaller than boulders.

cohesive strength—The cohesive strength of stonework pertains to one stone sticking to another and staying in place with the help of good mortar.

column—An ornamental or supporting pillar.

common bond—See American bond.

compass—An instrument or tool used to lay out work.

concrete—Hard, strong building material made of cement, aggregate, and water.

concrete finish—The surfacing of concrete by a cement mason.

conglomerate—Clastic sedimentary rock composed of rounded fragments cemented together by calcareous, siliceous, or other material.

consistency—Degree of density or viscosity.

construction units—Brick, tile, block, and stone used with or without mortar.

coping—Highest course of a wall of masonwork; usually slopes to carry off water.

corbel—One or more courses of units stepped upward and outward from a vertical surface, supporting a superincumbent weight.

corner—Place of directional change of the walls or surfaces of a structure, such as inside or outside corners.

cornice—Top course of a wall, projecting horizontally, generally topped by a blocking course to weight it down and hold it in place.

course—One horizontal layer of laid construction units.

course bed—Top of the last course laid upon which another course is to be placed.

cove—Recessed place in a structure.

cramps—Metal bars or irons bent at the ends to enter holes in construction units to hold them in place.

cross joint—Head joint between construction units in a course (true or variable) that connects the horizontal bed joints.

crowding the line—Calling hard laying; building an overhanging wall.

crown—Extrados or vertex of an arch.

cubic measure—Unit or series of units for measuring volume.

cull—An inferior construction unit.

culling—The sorting out of culls.

curing—Perfecting by chemical change; maintaining proper conditions of moisture and temperature. Having acquired its initial set and hardened, concrete is said to be in its curing stage.

curtain wall—Nonbearing wall not supported by girders or beams.

cut joint—Protruding mortar of cross and bed joints is cut off with a trowel.

darby—Long, narrow strip of wood with two handles; used as a plasterer's float.

depth—The distance from the face of the stone to the back of the stone in masonwork.

diagonal bond—Masonry bond in which headers are laid diagonally.

diagonal thrust—The weight load of a stone construction may cause diagonal thrust, which is weight load pressure, as well as other pressure through a wall. Pressure causes stonework in the wall to bulge out either in the back or in the face of the stonework.

diamond—Decorative brick design.

diaper—A pattern of brick, the designs of which connect with or grow out of one another.

dimension stone—Sized stone.

dipping—Dipping construction units into grout rather than spreading mortar on them before laying.

dowels—Anything used to hold construction units securely in place when incorporated into masonwork.

draft—Difference in pressure between outside and inside air that draws air through a chimney.

drafts—Grooves cut to a plane on stone by a chisel and hammer. Drafts pertain to such work as weathering, margin drafts, winding, etc.

dressing—Working the face of a stone to the required finish; also, squaring a stone for ashlar.

dry dash—Glitter, glass, or aggregate thrown on fresh plaster or stucco for a finish.

dry stone—Stone laid without mortar.

eastern method of troweling—Placing a trowel of mortar and immediately laying the brick.

edge set—Brick laid on edge showing the broad side of the brick.

efflorescence—Powder or crust formed on the masonry surface by the flow of water. Dampness brings salts to surface.

egg—Generally built for a sewer.

elliptical arch—Arch above an opening.

encaustic tile—Tile burned in various preparations to produce certain effects.

English bond—Masonry bond in which header courses alternate with stretched courses.

English cross bond—Modification of English bond in which stretcher courses break joints with each other.

extrados—Upper outer surface of an arch.

fabricated—A fabricated product is not a natural product. It is a manufactured product, such as a man-made stone or manufactured stone.

face—Outside surface of any construction unit that shows.

false screeds—Plaster or mortar laid or spread fair to a grade plane in strips 2″ to 3″ wide and 4′ to 6′ apart, of any length; used in place of screed strips as a ground when screeding or rodding to an even surface. Spot tile is used for the same purpose.

fat mortar—Mortar rich in a bonding material such as cement, lime, etc; having less aggregate.

filling-in—Laying the center of a wall.

fireclay—Clay that is refractory and able to withstand high temperatures without deforming; better grades contain at least 35 percent alumina when fired. It is used as a mortar where refractory material is required.

firebrick—Brick which will withstand more heat than regular brick.

fire stop—Masonwork placed between joists (as is brick nogging) to prevent fire from spreading.

flat arch—Arch with straight or almost straight horizontal intrados and extrados.

Flemish bone—Bond in which each course consists of headers and stretchers alternately laid so as to always break joints.

Flemish double bond—One header to two stretchers in a course.

Flemish triple bond—One header to three stretchers in a course.

flue—Inside passageway of a chimney stack for carrying off smoke.

flue liner—Flue tile.

flush—Bring a construction unit even (flush) with the surface of masonwork.

flushing—Slushing mortar into joints with a trowel.

flying buttress—Straight inclined masonry arch spanning an opening passageway to a solid pier; used to take up the thrust of brickwork.

footing—Substructure or bottom unit of a wall or column.

form—Shaped temporary holder that retains concrete or masonwork until it is set hard enough to sustain its own weight and hold its shape.

foundation—Supporting part of a wall or structure, usually below ground level and including footings.

frame high—Masonwork laid to the level of a framed opening; specifically, the top of a frame.

furring—Thin wood or metal pieces applied to joists, studs, or walls to form a level surface for lathing or plastering.

furring brick—Hollow brick large enough to bond. It is grooved to afford a key for plastering.

furring tile—Structural clay tile used for lining the inside of a wall. It carries no superimposed load.

gain ground—To speed up in laying or setting units of construction.

galleting—Pressing stone pebbles into the face mortar joint for strength or appearance.

galley—Roofed space open at the sides; also called an arcade or cloister.

garden wall—Any wall that stands without overload.

gauged putty—Putty stuff, each batch of which is mixed thin and in the same proportions to secure even setting.

gauged work—Forming, shaping, or proportioning of material as voussoirs in arches.

gingerbread work—Ornate masonry with mixed kinds of construction units.

glass block—Hollow, translucent block, usually with a ribbed exterior made by fusing two sections of clear pressed glass.

glazed tile—Ceramic tile.

glutinous masonry—All unnecessary material that will detract from the masonwork.

gothic arch—Pointed arch with a joint at its apex.

graded aggregate—Graded aggregate contains various particles of sand and gravel.

green brickwork—Brickwork in which the mortar has not yet set.

grounds—Wood or metal strips placed around all openings and along the top of the wall base to guide in finishing plaster; ground screeds.

grout—Mortar thin enough to pour.

hand-burned brick—Brick that received the proper amount of burning in the kiln.

haunch—That portion of an arch between skewback or springing and apex of an arch.

hawk—Small board or metal sheet with a handle on the underside; used to hold mortar.

header—The end of a brick laid to the face of brickwork.

header bond—Masonry bond in which all courses are header courses.

header course—A course of headers.

header high—The course below a header course.

head joint—Vertical masonry units joint between units in a course.

headway—Clearance below an arch.

heart bond—The meeting of two headers in the middle of a wall, with their joint covered by another header.

hearth—The floor of a fireplace and on which a fire may be built.

herringbond bond—Masonry bond in which construction units in adjacent rows slope slightly in reverse directions.

honed—A stone that is wrought smooth by sanding, grinding, or planing.

horseshoe arch—Arch above an opening.

hydrated lime—Slaked lime.

impervious—Not allowing entrance or passage through.

in-corners—Corners at the inside angle of the outside walls of a structure.

intitial set—The first taking up of water and achieving semihard condition.

insert—Setting or building of other units into the face of a wall or structure.

interlocking—Bonding construction units by lapping.

intrados—Interior curve of an arch.

inverted arch—Arch with the crown downward; used in foundations, sewers, and tunnels.

jack arch—A flat arch; also, a temporary, poorly constructed arch.

jamb—Side of an opening in a wall.

joggle—V-shaped sinking in the side of all the stone of the top course at the head or cross joints for the purpose of filling the openings with cement mortar to prevent shifting.

joint—Space between two adjacent units held together by mortar.

jointer—A tool used for joint work or jointing.

jointing—Finishing mortar joints with jointers.

Keene's cement—Hard finish gypsum plaster to which alum has been added.

keystone—Construction unit at the crown of an arch; the apex unit.

kneeler—Stone tailed well into the wall of the gable to resist the sliding tendency of the coping.

lacing course—Bonding the inside of unbonded stonework with brick or slabstone, as well as horizontal bands laid in bond rubble or cobble stonework.

lagging—Wooden strips for transferring to the centering form the weight of an arch under construction.

lateral thrust—Sidewise movement caused by load pressure. This can cause stone in a wall to slip out of place if not protected when laid.

lath—Groundwork on which plaster or stucco is spread.

laying overhand—Laying both faces of a wall when the scaffold is on only one side of a wall, necessitating reaching over to lay up the opposite side.

laying to bond—Keeping the bond plumb and true.

lead—Laying ahead or in advance of the line at the ends of the line, for the purpose of having lead units as a guide and a place to attach the line.

lean mortar—Mortar that lacks enough binder.

lewis—Dovetailed tenon that fits into a dovetail mortise in a stone to lift the stone.

lime—Calcium carbonate; a binder.

lime putty—Slaked lime in a putty state.

line—String stretched taut from lead course to lead course; used as a guide when laying construction units.

linear foot—One-foot distance along a straight line.

lintel—Horizontal architectural member spanning an opening and carrying the load above it.

lipped—Construction unit laid with its lower face edge protruding from the plumb face of the masonwork.

lock—Method of securing a laid unit.

longitudinal strength—Ability to withstand horizontal movement.

low gothic arch—A squatty Gothic arch.

lute—Wooden implement, resembling a rake without teeth; used to level off freshly poured concrete.

mantel—The ornamental shelf above a fireplace opening.

marginal line—The margin perpendicular outside line of the face of a structure.

measuring box—Box holding 1 cubic foot of material; used to proportion a mix.

models—Casts used to copy or to shape a plastic mix.

mortar—Plastic building material that hardens and binds.

mortar board—The board on which mortar is put for the mason.

mortar box—Mixing box.

mortar tub—Mixing tub.

muriatic acid—Commercial grade of hydrochloric acid; used as a masonry wash when diluted with water.

natural bed stone—Stone placed in a structure parallel to its stratification.

neat cement—Pure cement and water (no aggregate).

needling—Temporary support of needle beams (transverse floor beams).

niche—Recess in a wall.

nogging—Stonework filling open spaces of a wood frame as between rafters, joists, etc.

oakum—Loosely twisted fiber impregnated with a tar derivative; used to caulk seams and pack joints.

obtuse corner—Corner with more than a 90 degree angle and less than a 180 degree angle between its walls.

offset—A construction unit or course of masonry set back from the face of the unit or course on which it is set.

ogee arch—Arch in which the intrados form two contrasted ogee (or S) curves that meet in a point at the apex.

oriel—Semi-hexagonal bay window projecting from a wall and supported by a corbel.

out-corner—Corner outside of a structure, the flared joining walls of which are not more than 180 degrees apart.

outrigger—Projecting beam that supports a scaffold.

outside 4″—Face wall that is the thickness of its units' width.

overhand work—Laying brickwork (inside and outside faces) from one side of the wall.

overhang work—Wall face with corbeled cornice or outward battered wall.

overhead work—Work on ceilings or above the height of a mason.

panel—Sunken or raised section of a surface, set off by molding or other margin.

parapet—Low wall for protection or decoration; generally placed along the edge of a roof or terrace.

pargeting—Plastering the inside of the chimney flue and smoke chamber to give a smooth surface and help the draft; also, plastering the inside of a hollow wall such as waterproofing.

party wall—Wall or partition between two adjoining properties, with half its thickness on either property.

patching—Repairing.

peach basket—Temporary template used when building the head of a large chimney stack.

pebble—Small hard aggregate, or spall, placed in the mortar bed when laying stone. It holds the laid stone in place.

pebble dash—Rock dash.

peening—Dressing the face of a stone with a peen hammer.

perpend—Large unit extending through a wall as a binder and appearing on both sides.

piano basket—An outside swinging scaffold.

pick and dip—Eastern method.

pier—Vertical masonwork that supports the end of an arch, lintel, or beams.

pilaster—A structural pier which projects a third or less of its depth from the wall. It may or may not be load-bearing.

pitch—A slope from zero height at one end to inches or feet at the other end (e.g., pitch of a roof).

plaster—Mix whose binder's base is sulfate of lime, made from gypsum. Plaster is for interior use only, as it will not withstand moisture.

plastic—Capable of being modeled or shaped; a pliable mortar.

plinth—Square block serving as a base of a column, wall, or other structure.

plumb—Small weight (bob) attached to a line; used to indicate vertical direction.

plumb bond—Masonry bond in which corresponding joints of a bond are in vertical alignment.

plumb rule—Narrow board with plumb line and bob. A measuring rule divided into various size elevations of course heights considered as standard.

pointing—Filling in missed places in joints, striking or tooling the joint, or removing excess mortar with a frenchman or brush.

polishing—Smoothing, slicking, and shining the finished surface of any masonry unit or material

portland cement—Hydraulic cement made by finely pulverizing the clinker produced by calcining to incipient fusion a mixture of argillaceous and calcareous materials.

precast—An artificial unit; a mold, cast, etc., made before its use.

pressure—Pressure is exerted by weight load, whether as a wall load (dead or live) or lateral force.

pugging—Filling with sound-absorbing mortar.

punner—Tool used to ram, tamp, or consolidate.

put log—Plank holding cross support on a scaffold.

putty coat—Generally, the last coat applied to a plastered wall.

queen closer—Construction unit.

quoin—Corner construction unit.

rabbeting—Making a groove, principally in stone.

racking—Slight setting back of each course.

racking bond—Diagonal bond.

raggles—Grooves or raked-out joints for flashing.

rake—Remove mortar from a joint before it sets. An inclined face or raked end of a wall.

rangework—Ashlar laid in horizontal courses.

recess—Niche.

relieving arch—Arch over a lintel or incorporated in the wall; used to relieve or distribute weight of above wall.

reveal—A set-in at the jamb.

rich mix—Mortar with plenty of binder in it.

ring stone—Voussoir that shows.

riprap—Stone thrown together without order.

rise—Height of an arch from the middle of the spring line to the bottom of the key.

rock dash—Stucco finish in which crushed rock or pebbles are embedded.

rolled—Construction unit laid with the upper part of its face projecting beyond the plumb line of the wall.

roughly squared—Stone roughly squared with a hammer and laid, as in rubble stonework.

rough-pointing—Smearing mortar joints with the trowel.

rowlock—Course of brick laid on edge with the ends showing.

rubble—Unsquared stone or irregular size and shape.

rule—Measure used to measure distance from one point to another.

run—Walking space where masons work.

running bond—Masonry bond in which each unit overlaps units in adjoining courses.

saddle—Apex stone of a gable.

sand—Grains less than 2 millimeters in diameter, commonly of quartz; used as aggregate in mortar.

sand blasting—Engraving, cutting, or cleaning with a high-velocity stream of sand forcibly projected by air or steam.

sand cushion—Layer of sand separating two structural members, such as subbase and base slab for finished masonry of a floor.

scabbling—Knocking off only the roughest irregularities of a stone.

scaffold—Temporary, movable platform.

scaffold height—Raising the scaffold up to the work just done.

scale box—Box for measuring material.

scant—Batter-laid construction unit stepped inward from the plumb line.

scour—The best scouring tool to use when cleaning masonwork is a wire brush. Terrazzo floors need scouring on the finish coat with fine Carborundum rubbing stones.

scratch coat—First coat of plaster scratched to improve the bond with the next coat.

screed—A ground of wood, metal, or the material being used. It is placed to secure a plumb or level surface.

segmental arch—An arch above an opening.

selects—Construction units after culling.

semicircular arch—An arch above an opening.

semi-Gothic arch—An arch above an opening.

set—Solidified.

setting coat—Finish coat of plaster.

shell of a chimney—Outside wall of a chimney stack.

shores—Props or temporary underpinning.

shoved joint—Vertical joint made by shoving the bed mortar with the unit being laid. A push joint.

sill—Units at the bottom of an opening in a wall.

skew corbel—Projecting stone positioned at the bottom or lowest part of a gable for the purpose of not allaying the sliding of the units.

skewback—Course or courses of masonry with an inclined face against which voussoirs of an arch abut. A unit placed to receive pressure.

skintle—To set masonry units irregularly in a wall so they are out of line with the face of the wall ⅛″ to ¼″ or more.

slack lime—Is not putty lime.

slack line—Laying to a slack line makes an untrue course in the masonwork.

slaking—Treating lime with water, causing it to heat and crumble.

slope wall—A battered masonry wall out of plumb.

slump best—Test to determine the consistency of freshly mixed concrete by measuring the number of inches the mass settles after removal of a metal cone or cylindrical container open at the top and bottom. The drop in inches designates the "slump" of the mix, which tells the mason the number of days required for this mix to set hard for a 12, 21, or 28-day strength test. Little slump produces the best concrete.

slush joint—Grout of portland cement, sand, and water. Thin mortar slushed into a joint.

smoke chamber—Part of a fireplace, between top of the throat to bottom of the flue.

smoke shelf—Baffle in a flue designed to return downdraft out the top of a chimney stack.

smooth-pointed—A slicked joint.

sneck—Small roughly squared stone.

soffit—Underside of a part of a ceiling, overhang, or cornice; intrados of an arch.

soldier course—Masonry course in which the units are laid with their end placed on the mortar bed.

spalls—Fragments broken from a stone and having at least one featheredge.

span—Distance of the opening between jambs, abutments, etc.

spandrel—Area between extrados of an arch and adjacent moldings or adjoining arch. The space between haunch span and right angle.

spirit plumb—Level.

splay—Angle, slope, or bend from something, such as the splay from the jamb of a window or door that increases the amount of light through the opening.

split—Stretcher halved horizontally.

spreading mortar—This is carried on after the mortar is on the mortar bed for the purpose of spreading the mortar. A trowel is used to move the mortar to the edge of the mortar bed.

spring course—The spring course is the course from which an arch springs.

spring line—Imaginary line connecting the two opposite points at which the curve of an arch begins.

springer—Unit at the impost of an arch. The lowest voussoir of an arch.

spudding—Punning.

square—Builder's unit equal to 100 square feet.

square corner—90 degree angle corner.

squared—Salient angle or arris of a 90 degree angle presents a squared appearance to a stone, whether the stone has an oblong or square face.

squint—Brick cut to an oblique angle.

stack bond—Masonry units stacked one on top of the other, having the same vertical joints.

stanchions—Construction units forming the inside angle of a jamb; generally made of concrete, which is less costly than dressing or shaping stone.

stippling—Textured finish for plaster or stucco.

stock—Construction units, mortar, etc., used in masonwork.

stopped wall—Terminal height of a wall.

story high—From one floor to the next floor is considered a story.

story pole—Pole cut to the proposed clear height between finished floor and ceiling marked with dimensions for courses, sills, frames, string trimmings, etc.

straight arch—Flat arch.

straightedge—Screeding tool or rod of wood or metal; used to secure a grade or straight surface.

stretcher—Brick laid with its length parallel to the face of the wall.

stretcher bond—Masonry bond in which all units are laid as stretchers breaking joints.

string course—Belt course.

stringing mortar—A trowelful of mortar is strung out on a mortar bed.

struck joint—Joint in which green mortar is recessed with the trowel.

structural steel—Steel of various sizes and lengths used in framework construction.

stucco—Material (usually portland cement, sand, and a small percentage of lime) applied in a plastic state to form a hard covering for exterior surfaces. A plaster that resists moisture.

stuff—Plaster, mortar, grout, and cement mix.

sunk work—Incised face finish of wrought stone

superstructure—Structure built as a vertical extension of a structure.

support unit—Skewback.

tape—A measuring tool used to measure from point to point.

tapping—Pounding a brick to place with a trowel handle or blade.

tempering—Addition of water to a mix after it has nearly reached or is nearly in its initial set.

template—Gauge or form used as a pattern or copy to follow.

tender—The laborer that tends a bricklayer—called hodman or hod carrier.

tensile strength—Pertains to the staying together of brickwork. Principally, it is the strength stone or mortar has, and the ability to hold together and not pull apart.

terra-cotta—Glazed or unglazed ceramic building material.

terrazzo—Mosiac surface made by embedding small pieces of marble in mortar and, after hardening, grinding and polishing the surface.

texture—Applied to a brick's face appearance.

thickness of wall—Distance from face to face through a wall.

three-quarter bat—A part of a brick, three-fourth size in length.

throat—Opening in a fireplace from the top of the firebox to the smoke chamber.

throating—Groove on the underside of sills, copings, etc., that prevents water from running back into the wall.

through bond—Transverse bond formed by a member that extends crosswise through a wall.

through wall—Perpend that extends from face to face through a wall.

throwing mortar—Filling up cracks by slushing mortar from the trowel.

tier—Vertical layer of brickwork whose thickness is the width of a brick.

tooled—Dressed.

tools—Necessary implements when building masonwork.

toothed—Leaving masonwork perpendicular in and out for future continuation of the masonwork.

toothing—Alternate courses of brick projecting at the end of a wall to permit bonding into a later continuation of the wall.

toothing-in—Joining masonwork to a toothed wall.

topping—Finish coat or pour.

trammel—A device for drawing ellipses.

transverse joint—Joint extending through a wall.

transverse strength—Ability to withstand pressure.

trig—Brick bedded to the proper height to hold a mason's line level in the center of a course.

trimmer—Beam that receives a floor framing header to keep floor joists away from a chimney.

trimmer arch—Arch built between trimmers in the thickness of an upper floor to support a hearth.

tucking—Filling mortar joints after masonry units are laid.

tuck-pointing—Finishing mortar joints with a narrow ridge of putty or fine lime mortar. In tuck-pointing, the old mortar in the joints is routed out and replaced with new mortar nearly the same color as the construction units in the wall. A grooving tool or jointer is used to indent the newer mortar. This groove is filled with a different color mortar. When the mortar has set sufficiently, it is cut off and trimmed with a trowel or frenchman.

tying in—Joining one wall to another.

typanum—Pedimental space between intrados of a relieving arch and above the lintel or spring line.

vermiculated—Stone dress appearing to be covered with worm tracks.

vertical joint—Cross joint.

vertical strength—Capability of holding a load from above.

vibrating—Method of settling an agglomerated mass into place.

vitrification—Fused state by burning.

voids—Air spaces in material.

voussoir—Whole or wedge-shaped units of an arch.

wainscot—Lower 3' or 4' of an interior wall when finished differently from the rest of the wall.

wall plate—First member above a wall and on which the roof is laid.

wall ties—Mortar-embedded ties that hold walls or units together.

wash—Slope given stone to shed water.

washing—Raised tooled mortar joint, generally a floor joint, which will drain off water.

water table—Course near the base of a structure, projecting to throw off water.

weathering—Decay or deterioration by the effects of the elements of nature.

web—Partitions forming cells in tile or block.

weep hole—Hole left mortar-free in masonry that will drain off water.

western method of troweling—Stringing mortar on the bed, spreading it toward the outer edges, and cutting off the runover with a trowel before laying the masonwork.

wind shelf—Smoke shelf.

wing wall—Extended abutment to retain earth.

withe—The partition between flues in the same chimney.

wrought—The shaping of a stone to get it ready for setting.

Appendix A

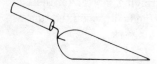

Tables of Weights and Measures

There are two dominant systems used to measure quantity: *United States* and *metric*. The U.S. system is employed in all measurements in this book. Should the need arise, a change to the metric system is possible with the use of the following information and tables.

The U.S. system listed in Tables A-1 through A-3 shows the units of measure regarding length, capacity, and weight. Each unit has its own measure of value. Such value may be changed to metric equivalents (Tables A-4 through A-9). Measures and weights of materials are given in Tables A-10 and A-11. Table A-12 lists circular measures.

TEMPERATURE CONVERSIONS

You must watch the temperature. Mortars should not freeze or stonework is worthless. Since there are two temperature thermometers, *Fahrenheit* and *centigrade*, it may be necessary to convert the scale reading from one to the other.

Table A-1. Units of Distance Measure in the
United States System and Their Abbreviations.

U.S. SYSTEM: DESIGNATIONS AND ABBREVIATIONS		
Unit Name	Area (Square)	Quantity (Cubic)
inch (in.)	square inch (sq. in.)	cubic inch (cu. in.)
foot (ft.)	square foot (sq. ft.)	cubic foot (cu. ft.)
yard (yd.)	square yard (sq. yd.)	cubic yard (cu. yd.)
rod (rd.)	square rod (sq. rd.)	cubic rod (cu. rd.)
mile (mi.)	square mile (sq. mi.)	cubic mile (cu. mi.)

Table A-2. Units of Capacity Measure in the American System and Their Abbreviations.

CAPACITY UNIT MEASURES		
gill (gi.) pt.	quart (qt.) peck (pk.)	gallon (gal.) bushel (bu.)

Table A-3. Units of Weight Measure in the United States System and Their Abbreviations.

WEIGHT UNIT MEASURES		
grain (gr.) pound (lb.) hundredweight (cwt.)	dram (dr.) fluid oz. (fl. oz.) short ton (s. t.)	ounce (oz.) dry oz. (dry oz.) long ton (l. t.)

Table A-4. Units of Length, Capacity and Weight Measure in the Metric System, Together with Abbreviations.

METRIC ABBREVIATIONS AND SYMBOLS	
Length Measure	**Square Measure**
millimeter (mm.) centimeter (cm.) decimeter (dm.) meter (m.) decimeter (dkm.) hectometer (hm.) kilometer (km.)	square millimeter (sq. mm.) square centimeter (sq. cm.) square decimeter (sq. dm.) square meter (sq. m.) square decimeter (sq. dkm.) square hectometer (sq. hm.) square kilometer (sq. km.)

Cubic Measure	**Capacity**
cubic millimeter (cu. mm.) cubic centimeter (cu. cm.) cubic decimeter (cu. dm.) cubic meter (cu. m.) cubic decameter (cu. dkm.) cubic hectometer (cu. hm.) cubic kilometer (cu. km.)	milliliter (ml.) centiliter (cl.) deciliter (dl.) liter (l.) decaliter (dkl.) hectoliter (hl.) kiloliter (kl.)

Weight Measure	**Weight Measure**
milligram (mg.) centigram (cg.) decigram (dg.) gram (g.)	decagram (dkg.) hectogram (hg.) kilogram (kg.) metric ton (M.T.)

Table A-5. American Units of Length Measure Changed to Metric Equivalents.

U.S. UNITS OF MEASURE CHANGED TO METRIC EQUIVALENTS		
Units of Length	U.S. System	Metric System
inch (in.)	0.083 ft.	25.4 mm.; 2.54 cm.
foot (ft.)	12 in.⅓ yd.	30.48 cm.; 0.3048 m.
yard (yd.)	36 in.; 3 ft.	0.9144 m.
rod (rd.)	16½ ft.; 5 ½ yd.	5.0292 m.
land mile (mi.)	1760 yd.; 5280 ft.	1609.344 m.
Nautical mile	1.151 land mile	1.852 km.
sq. in.	0.007 sq. ft.	6.4516 sq. cm.
sq. ft.	144 sq. in.	929.030 sq. cm.
sq. yd.	1296 sq. in.; 9 sq. ft.	0.8361 sq. m.
sq. acre; (sq. A.)	43,560 sq. ft.	4.047 sq. m.
sq. mi.	640 acres	2.590 sq. km.
cu. in.	0.00058 cu. ft.	16.387 cu. cm.
cu. ft.	1728 cu. in.	0.028 cu. m.
cu. yd.	27 cu. ft.	0.765 cu. m.

Table A-6. United States Units of Weight Measure and Their Metric Equivalents.

WEIGHT MEASURE		
Units of Weight	U.S. System	Metric
gr.	0.036 dr.; 0.002285 oz.	64.79891 mg.
dr.	27.344 gr.; 0.0625 oz.	1.772 gr.
oz.	16 dr.; 437.5 gr.	28.350 gr.
lb.	16 oz.; 7000 gr.	453.59237 gr.
s. t.	2000 lb.	0.907 M.T.
l. t.	1.12 s.t.; 2240 lb.	1.016 M.T.

Table A-7. American Units of Capacity Measure and Metric Equivalents.

CAPACITY MEASURE		
Units of Capacity	U.S. System	Metric
f. oz. (liquid)	8 fl. dr.; 1.804 cu. in.	29.573 ml.
pt. ″	1 fl. oz.; 528.875 cu. in.	0.473 l.
qt. ″	2 pt.; 57.75 cu. in.	0.946 l.
gal. ″	4 qt.; 231 cu. in.	3.785 l.
bbl. ″	31 to 42 gal. (180 lb. U.S.)	81.65 kg.
pt. dry	½ qt.; 33.6 cu. in.	0.551 l.
qt. ″	2 pt.; 67.2 cu. in.	1.101 l.
pk. ″	8 qt.; 537.606 cu. in	8.810 l.
bu. ″	4 pk.; 2150.42 cu. in.	35.2381 l.

Table A-8. British Units of Liquid and Dry Capacity Measure.

oz.	0.961 U.S. fl. oz. 1.734 cu. in.	28.413 ml.
pt.	1.032 U.S. dry pt. 1.201 U.S. fl. pt. 34.678 cu. in.	568.26 ml.
qt.	1.032 U.S. dry qt. 1.201 U.S. fl. qt.	1.136 l.
pk.	554.84 cu. in.	0.009 cu. m.
bu.	1.032 U.S. bu. 2.219.36 cu. in.	0.036 m.

Table A-9. Metric Units of Measure and Their American Equivalents.

mm.	0.03937 in.	0.00328 ft.
cm.	0.3937 in.	0.0328 ft.
dm.	3.937 in.	0.328 ft.
m.	39.37 in.	3.28 ft.
dkm.	393.7 in.	32.8 ft.
dm.	3937 in.	328 ft.
km.	0.621 mi.	3280.8 ft.

Table A-10. Measures and Weights of Materials.

MEASURES AND WEIGHTS OF MATERIAL SUPPLIES (Approximately)		
1 cu. ft.	water	62.4 lbs.
1 gal.	muriatic acid	9+ lbs.
1 bucket (12 qt.)	lime putty	30 lbs.
1 bucket (12 qt.)	hydrated lime	16 lbs.
1 bucket (12 qt.)	Keene's cement	30 lbs.
1 bucket (12 qt.)	sand	40 lbs.
1 bushel (4 pk.) bu.	portland cement	126 lbs.
1 bu.	slaked quicklime putty	100 lbs.
1 bu.	hydrated lime putty	109 lbs.
1 bu.	hydrated lime	50 lbs.
1 bu.	quicklime	72 lbs.
1 bu.	Keene's cement	94 lbs.
1 T.	clay	16 cu. ft.
1 T.	sand	18 to 22 cu. ft.
1 cu. yd.	sand (dry)	2500 to 2700 lbs.
1 bag (1 cu. ft.)	portland cement	94 lbs.
1 bag	hydrated lime	50 lbs.
1 bbl.	cement	376 lbs.
1 bbl.	quicklime	180 lbs.
1 cu. ft.	hydrated lime	40 lbs.
1 cu. ft.	lime	45 lbs.
1 cu. ft.	hydrated lime putty	83 lbs.
1 cu. ft.	lime putty	2.7 bucket (12 qt.)

Table A-11. Cubic Foot Weight of Selected Materials.

loose dirt .... 90 to 100 lbs.	limestone .... 170 to 184 lbs.
clay and gravel .... 300 lbs.	random and split
sand .... 90 to 113 lbs.	ashlar stone .... 160 lbs.
new mortar .... 115 lbs.	sandstone .... 140 to 144 lbs.
old mortar .... 90 lbs.	granite .... 160 to 172 lbs.
quicklime .... 58 lbs.	marble .... 168 to 172 lbs.
cement mortar .... 112 lbs.	loadstone .... 305 lbs.
portland cement .... 94 lbs.	slate .... 166 lbs.
hydrated lime .... 40 lbs.	rubble .... 130 lbs.
gypsum .... 143 lbs.	common stone .... 158 lbs.
tar .... 64 lbs.	quartz .... 166 lbs.
asbestos .... 188 lbs.	prophyr .... 172 lbs.
cork .... 15 lbs.	portland stone .... 157 lbs.
asphalt .... 103 lbs.	porcelain stone .... 159 lbs.
pumice block .... 103 lbs.	brick .... 119 to 128 lbs.
	concrete block .... 48 lbs.

(The blocks are one-tenth short when laying a square foot).

1 minute (')	60 seconds (")
1 degree (°)	60', 1 .°
1 circle	360°

Table A-12. Units of Circular Measure.

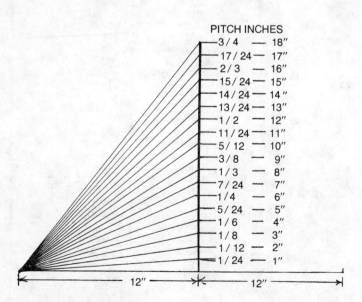

Fig. A-1. This graph illustrates pitch.

To convert the Fahrenheit reading to the centigrade, take 32 from the Fahrenheit reading and multiply by five-ninths. The answer is the centigrade temperature reading.

To convert the centigrade reading to the Fahrenheit temperature reading, multiply the centigrade reading by five-ninths and add 32. The answer is the Fahrenheit reading.

PITCH

The slopes of roofs, wall tops, etc., are figured in pitches. Using 24′ as the base (width), divide the pitch desired into 24. The answer will be the rise in inches. Using 24 feet as the base (width), divide the inches of the rise desired into 24. The answer is the pitch. See Fig. A-1.

Appendix B

Concrete, Cement, and Nails Information

Table B-1. Recommended Maximum Permissible Water-Cement Ratios for Different Types of Structures and Degrees of Exposure.

Type of structure	Exposure conditions**					
	Severe wide range in temperature or frequent alternations of freezing and thawing (air-entrained concrete only) (gallons/sack)			Mild temperature rarely below freezing, or rainy, or arid (gallons/sack)		
	In air	At water line or within range of fluctuating water level or spray		In air	At water line or within range of fluctuating water level or spray	
		In fresh water	In sea water or in contact with sulfates†		In fresh water	In sea water or in contact with sulfates†
A. Thin sections such as reinforced piles and pipe	5.5	5	4.5	6	5.5	4.5
B. Bridge decks	5	5	4.5	5.5	5.5	5
C. Thin sections such as railings, curbs, sills, ledges, ornamental or architectural concrete, and all sections with less than 1-in. concrete cover over reinforcement	5.5	....	....	6	5.5	....
D. Moderate sections, such as retaining walls, abutments, piers, girders, beams	6	5.5	5	††	6	5
E. Exterior portions of heavy (mass) sections	6.5	5.5	5	††	6	5

F. Concrete deposited by tremie under water .... 5 .... 5 .... 5 5

G. Concrete slabs laid on the ground .... 5.5 .... †† ....

H. Pavements .... 6 .... 6 ....

I. Concrete protected from the weather, interiors of buildings, concrete below ground .... †† .... †† ....

J. Concrete which will later be protected by enclosure or backfill but which may be exposed to freezing and thawing for several years before such protection is offered .... .6 .... †† 5

*Adapted from Recommended Practice for Selecting Proportions for Concrete (ACI 613-54).

**Air-entrained concrete should be used under all conditions involving severe exposure and may be used under mild exposure conditions to improve workability of the mixture.

†Soil or groundwater containing sulfate concentrations of more than 0.2 per cent. For moderate sulfate resistance, the tricalcium aluminate content of the cement should be limited to 8 per cent, and for high sulfate resistance to 5 percent. At equal cement contents, air-entrained concrete is significantly more resistant to sulfate attack than non-air-entrained concrete.

††Water-cement ratio should be selected on basis of strength and workability requirements, but minimum cement content should not be less than 470 lbs. per cubic yard.

Table B-2. Age and Compressive Strength
Relationship for Types I and III Air-Entrained Portland Cement.

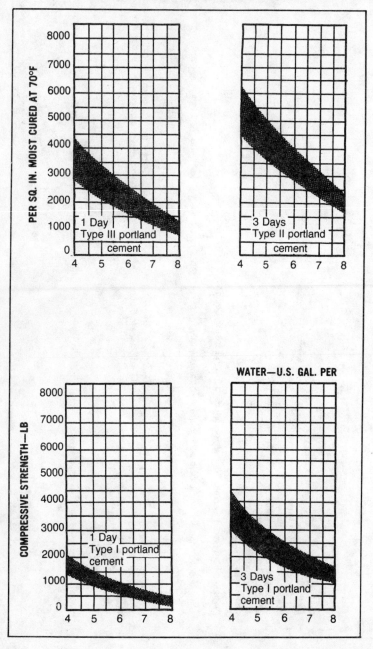

Table B-2. Age and Compressive
Strength Relationship for Types I and III
Air-Entrained Portland Cement (continued from page 234).

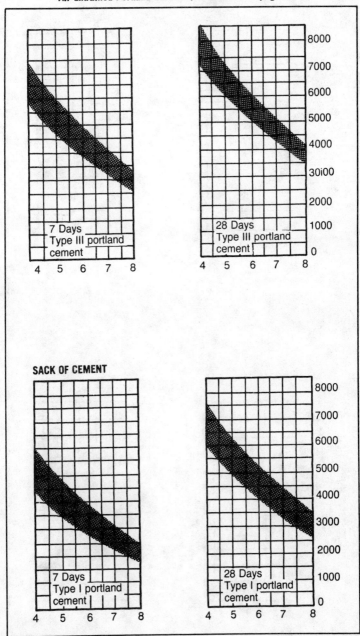

SACK OF CEMENT

7 Days
Type III portland
cement

28 Days
Type III portland
cement

7 Days
Type I portland
cement

28 Days
Type I portland
cement

Table B-2. Age and Compressive
Strength Relationship for Types I and III
Air-Entrained Portland Cement (continued from page 235).

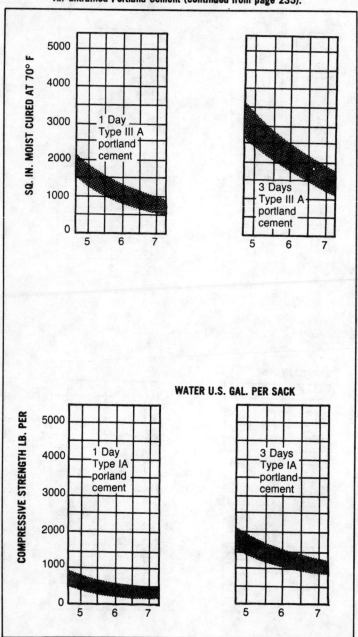

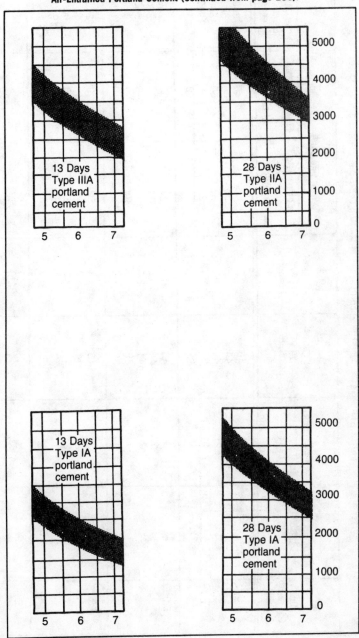

Table B-2. Age and Compressive Strength Relationship for Types I and III Air-Entrained Portland Cement (continued from page 236).

13 Days Type IIIA portland cement

28 Days Type IIA portland cement

13 Days Type IA portland cement

28 Days Type IA portland cement

Table B-3. Suggested Trial Mixes for Non-Entrained Concrete of Medium Consistency with a 3 to 4" Slump.

Water-cement ratio Gal per sack	Maximum size of aggregate inches	Air content (entrapped air) per cent	Water gal per cu yd of concrete	Cement sacks per cu yd of concrete	With fine sand—fineness modulus = 2.50		
					Fine aggregate per cent of total aggregate	Fine aggregate lb per cu yd of concrete	Coarse aggregate lb per cu yd of concrete
4.5	3/8	3	46	10.3	50	1240	1260
	1/2	2.5	44	9.8	42	1100	1520
	3/4	2	41	9.1	35	960	1800
	1	1.5	39	8.7	32	910	1940
	1½	1	36	8.0	29	880	2110
5.0	3/8	3	46	9.2	51	1330	1260
	1/2	2.5	44	8.8	44	1130	1520
	3/4	2	41	8.2	37	1040	1800
	1	1.5	39	7.8	34	990	1940
	1½	1	36	7.2	31	960	2110
5.5	3/8	3	46	8.4	52	1390	1260
	1/2	2.5	44	8.0	45	1240	1520
	3/4	2	41	7.5	38	1090	1800
	1	1.5	39	7.1	35	1040	1940
	1½	1	36	6.5	32	1000	2110
6.0	3/8	3	46	7.7	53	1440	1260
	1/2	2.5	44	7.3	46	1290	1520
	3/4	2	41	6.8	39	1130	1800
	1	1.5	39	6.5	36	1080	1940
	1½	1	36		32	1040	2110

6.5	3/8	3	46	7.1	46	54	1480	1260
	1/2	2.5	44	6.8	44	46	1320	1520
	3/4	2	41	6.3	41	39	1190	1800
	1	1.5	39	6.0	39	37	1120	1940
	1½	1	36	5.5	36	34	1070	2110
7.0	3/8	3	46	6.6	46	55	1520	1260
	1/2	2.5	44	6.3	44	47	1360	1520
	3/4	2	41	5.9	41	40	1200	1800
	1	1.5	39	5.6	39	37	1150	1940
	1½	1	36	5.1	36	34	1100	2110
7.5	3/8	3	46	6.1	46	55	1560	1260
	1/2	2.5	44	5.9	44	48	1400	1520
	3/4	2	41	5.5	41	41	1240	1800
	1	1.5	39	5.2	39	38	1190	1940
	1½	1	36	4.8	36	35	1130	2110
8.0	3/8	3	46	5.7	46	56	1600	1260
	1/2	2.5	44	5.5	44	48	1440	1520
	3/4	2	41	5.1	41	42	1230	1800
	1	1.5	39	4.9	39	39	1220	1940
	1½	1	36	4.5	36	35	1160	2110

Table B-3. Suggested Trial Mixes for Non-Entrained Concrete of Medium Consistency with a 3 to 4" Slump (continued from page 238).

Water-cement ratio Gal per sack	With average sand—fineness modulus = 2.75			With coarse sand—fineness modulus = 2.90		
	Fine aggregate percent of total aggregate	Fine aggregate lb per cu yd of concrete	Coarse aggregate lb per cu yd of concrete	Fine aggregate percent of total aggregate	Fine aggregate lb per cu yd of concrete	Coarse aggregate lb per cu yd of concrete
4.5	52	1310	1190	54	1350	1150
	45	1170	1450	47	1220	1400
	37	1030	1730	39	1080	1680
	34	980	1870	36	1020	1830
	32	960	2030	33	1000	1990
5.0	54	1400	1190	56	1440	1150
	46	1250	1450	48	1300	1400
	39	1110	1730	41	1160	1680
	36	1060	1870	38	1100	1830
	34	1040	2030	35	1080	1990
5.5	55	1460	1190	57	1500	
	47	1310	1450	49	1360	
	40	1160	1730	42	1210	
	37	1110	1870	39	1150	
	35	1080	2030	36	1120	
6.0	56	1510	1190	57	1550	1150
	48	1360	1450	50	1410	1400
	41	1200	1730	43	1250	1600
	38	1150	1870	39	1190	1830
	36	1120	2030	37	1160	1990

6.5	57	1550	1190	58	1590	1150
	49	1390	1450	51	1440	1400
	42	1240	1730	43	1290	1680
	39	1190	1870	40	1230	1830
	36	1150	2030	37	1190	1990
7.0	57	1590	1190	59	1630	1150
	50	1430	1450	51	1480	1400
	42	1270	1730	44	1320	1680
	39	1220	1870	41	1260	1830
	37	1180	2030	38	1220	1990
7.5	58	1630	1190	59	1670	1150
	50	1470	1450	52	1520	1400
	43	1310	1730	45	1370	1600
	40	1260	1870	42	1300	1830
	37	1210	2030	39	1250	1990
8.0	58	1670	1190	60	1710	1150
	51	1520	1450	53	1560	1400
	44	1350	1730	45	1400	1680
	41	1290	1870	42	1330	1830
	38	1250	2030	39	1280	1990

*Increase or decrease water per cubic yard by 3 percent for each increase or decrease of 1″ in slump, then calculate quantities by absolute volume method. For manufactured fine aggregate, increase percentage of fine aggregate by 3 and water by 17 lb. per cubic yard of concrete. For less workable concrete, as in pavements decrease percentage of fine aggregate by 3 and water by 8 lb. per cubic yard of concrete.

241

Table B-4. Trial Mixes for Air-Entrained Concrete of Medium Consistency with a 3 to 4" Slump.

Water-cement ratio Gal per sack	Maximum size of aggregate inches	Air Content (entrapped air) percent	Water gal per cu yd of concrete	Cement sacks per cu yd of concrete	With fine sand—fineness modulus = 2.50		
					Fine aggregate per cent of total aggregate	Fine aggregate lb per cu yd of concrete	Coarse aggregate lb per cu yd of concrete
4.5	3/8	7.5	41	9.1	50	1250	1260
	1/2	7.5	39	8.7	41	1060	1520
	3/4	6	36	8.0	35	970	1800
	1	6	34	7.8	32	900	1940
	1 1/2	5	32	7.1	29	870	2110
5.0	3/8	7.5	41	8.2	51	1330	1260
	1/2	7.5	39	7.8	43	1140	1520
	3/4	6	36	7.2	37	1040	1800
	1	6	34	6.8	33	970	1940
	1 1/2	5	32	6.4	31	930	2110
5.5	3/8	7.5	41	7.5	52	1390	1260
	1/2	7.5	39	7.1	44	1190	1520
	3/4	6	36	6.5	38	1090	1800
	1	6	34	6.2	34	1010	1940
	1 1/2	5	32	5.8	32	970	2110
6.0	3/8	7.5	41	6.8	53	1430	1260
	1/2	7.5	39	6.5	45	1230	1520
	3/4	6	36	6.0	38	1120	1800
	1	6	34	5.7	35	1040	1940
	1 1/2	5	32	5.3	32	1010	2110

6.5	3/8	7.5	41	6.3	54	1460	1260
	1/2	7.5	39	6.0	45	1260	1520
	1/2	6	36	5.5	39	1150	1800
	1	6	34	5.2	36	1080	1940
	1½	5	32	4.9	33	1040	2110
7.5	3/8	7.5	41	5.9	54	1500	1260
	1/2	7.5	39	5.6	46	1300	1520
	3/4	6	36	5.1	40	1180	1800
	1	6	34	4.9	36	1100	1940
	1½	5	32	4.6	33	1060	2110
7.0	3/8	7.5	41	5.5	55	1530	1260
	1/2	7.5	39	5.2	47	1330	1520
	3/4	6	36	4.8	40	1210	1800
	1	6	34	4.5	37	1140	1940
	1½	5	32	4.3	34	1090	2110
8.0	3/8	7.5	41	5.1	55	1560	1260
	1/2	7.5	39	4.9	47	1360	1520
	3/4	6	36	4.5	41	1240	1800
	1	6	34	4.3	37	1160	1940
	1½	5	32	4.0	34	1110	2110

Table B-4. Trial Mixes for Air-Entrained Concrete of Medium Consistency with a 3 to 4" Slump (continued from page 242).

Water-cement ratio Gal per sack	With average sand—fineness modulus = 2.75			With coarse sand—fineness modulus = 2.90		
	Fine aggregate percent of total aggregate	Fine aggregate lb per cu yd of concrete	Coarse aggregate lb per cu yd of concrete	Fine aggregate percent of total aggregate	Fine aggregate lb per cu yd of concrete	Coarse aggregate lb per cu yd of concrete
4.5	53	1320	1190	54	1360	1150
	44	1130	1450	46	1180	1400
	38	1040	1730	39	1090	1680
	34	970	1870	36	1010	1830
	32	950	2030	33	990	1990
5.0	54	1400	1190	56	1440	1150
	46	1210	1450	47	1260	1400
	39	1110	1730	41	1160	1630
	36	1040	1870	37	1080	1830
	33	1010	2030	35	1050	1990
5.5	55	1460	1190	57	1500	1150
	46	1260	1450	48	1310	1400
	40	1160	1730	42	1210	1680
	37	1080	1870	38	1120	1830
	34	1050	2030	35	1090	1990
6.0	56	1500	1190	57	1540	1150
	47	1300	1450	49	1350	1400
	41	1190	1730	42	1240	1680
	37	1110	1870	39	1150	1830
	35	1090	2030	36	1130	1990

6.5	56	1530	1190	58	1570	1150
	48	1330	1450	50	1380	1400
	41	1220	1730	43	1270	1680
	38	1150	1870	39	1190	1830
	36	1120	2030	37	1160	1990
7.0	57	1570	1190	58	1610	1150
	49	1370	1450	50	1420	1400
	42	1250	1730	44	1300	1680
	38	1170	1870	40	1210	1830
	36	1140	2030	37	1180	1990
7.5	57	1600	1190	59	1640	1150
	49	1400	1450	51	1450	1400
	43	1280	1730	44	1330	1680
	39	1210	1870	41	1250	1830
	37	1170	2030	38	1210	1990
8.0	58	1630	1190	59	1670	1150
	50	1430	1450	51	1480	1400
	43	1310	1730	44	1360	1680
	40	1230	1870	41	1270	1830
	37	1190	2030	38	1230	1990

*Increase or decrease water per cubic yard by 3 percent for each increase of 1″ in slump, then calculate quantities by absolute volume method For manufactured fine aggregate, increase percentage of fine aggregate by 3 and water by 17 lb. per cubic yard of concrete. For less workable concrete, as in pavements decrease percentage of fine aggregate by 3 and water by 8 lb. per cubic yard of concrete.

Table B-5. Approximate Mixing Water Requirements for Different Slumps and Maximum Sizes of Aggregates.

Maximum size of aggregate, in.	Recommended average total air content, percent×	Air-entrained concrete			Approximate amount of entrapped air, percent	Non-air-entrained concrete		
		Slump in.				Slump, in.		
		1 to 2	3 to 4	5 to 6		1 to 2	3 to 4	5 to 6
		Water, gal. per cu. yd. of concrete**				Water, gal. per cu. yd. of concrete**		
3/8	7.5	37	41	43	3.0	42	46	49
1/2	7.5	36	39	41	2.5	40	44	46
3/4	6.0	33	36	38	2.0	37	41	43
1	6.0	31	34	36	1.5	36	39	41
1½	5.0	29	32	34	1.0	33	36	38
2	5.0	27	30	32	0.5	31	34	36
3	4.0	25	28	30	0.3	29	32	34
6	3.0	22	24	26	0.2	25	28	30

*Adapted from Recommended Practice for Selecting Proportions for Concrete (ACI 613-54).

**These quantities of mixing water are for use in computing cement factors for trial batches. They are maximums for reasonably well-shaped angular coarse aggregates graded within limits of accepted specifications.

†Plus or minus 1 per cent.

Table B-6. Details of Common Nails.

PENNY SIZE	LENGTH (INCHES)	GAUGE	NUMBER PER POUND
2	1	155	840
3	1¼	14	540
4	1½	12½	300
6	2	11½	160
8	2½	10¼	100
10	3	9	65
12	3¼	9	65
16	3½	8	45
20	4	6	30
30	4½	5	20
40	5	4	17
50	5¼	3	14
60	6	2	11

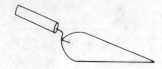

Index

Edited by Robert Ostrander